Vanderson Vieira Batista

Agronomic studies

Vanderson Vieira Batista

Agronomic studies

Scientific research

ScienciaScripts

Imprint

Cover image: www.ingimage.com

This book is a translation from the original published under ISBN 978-613-9-66942-4.

Publisher:
Sciencia Scripts
is a trademark of
Dodo Books Indian Ocean Ltd. and OmniScriptum S.R.L publishing group

120 High Road, East Finchley, London, N2 9ED, United Kingdom
Str. Armeneasca 28/1, office 1, Chisinau MD-2012, Republic of Moldova, Europe
Printed at: see last page
ISBN: 978-620-8-11566-1

Ana Claudia Schllemer dos Santos
Master's student in Agroecosystems
Federal Technological University of Paraná (UTFPR) - Dois Vizinhos/PR
Forestry Engineer
anaschllemer@gmail.com

Cleverson Alessio da Silva
Master's student in Regional Development
Federal Technological University of Paraná (UTFPR) - Pato Branco/PR
Management Processes Technologist
Agricultural Technician
scleverson8@gmail.com

Cleverson Luiz Giacomel
Agronomy student
Federal Technological University of Paraná (UTFPR) - Dois Vizinhos/PR
cleverson_giacomel@hotmail.com

Eduardo Júnior de Oliveira da Silva
Agronomy student
Federal Technological University of Paraná (UTFPR) - Dois Vizinhos/PR
duduardodasilva50@gmail.com

Karine Fuschter Oligini
Master's student in Agronomy
Federal Technological University of Paraná (UTFPR) - Pato Branco/PR
Agronomist
karine_oligini@hotmail.com

Lucas Gonçalves de Oliveira
Agronomy student
Federal Technological University of Paraná (UTFPR) - Dois Vizinhos/PR
lucasgoncalvesoliveira44@gmail.com
Lucas Link
Master's student in Agroecosystems
Federal Technological University of Paraná (UTFPR) - Dois Vizinhos/PR
Agricultural Engineer
lucas_tlink@hotmail.com

Marciani Balbinotti France
Zootechnics student
Federal Technological University of Paraná (UTFPR) - Dois Vizinhos/PR
marciani@alunos.utfpr.edu.br

Paulo Fernando Adami
Dr Professor of Agronomy
Federal Technological University of Paraná (UTFPR) - Dois Vizinhos/PR
Agricultural Engineer
pauloadami@utfpr.edu.com

Paulo Roberto Rabelo
Agronomy student
Federal Technological University of Paraná (UTFPR) - Dois Vizinhos/PR
paulo.rb@hotmail.com

Roniel Giaretta
Postgraduate Diploma in Fallow Crop Management
Federal Technological University of Paraná (UTFPR) - Dois Vizinhos/PR
Agricultural Engineer
ronielgiaretta@hotmail.com

Vanderson Vieira Batista
Master's student in Agroecosystems
Federal Technological University of Paraná (UTFPR) - Dois Vizinhos/PR
Agricultural Engineer
vandersonvbatista@hotmail.com

Summary

CHAPTER 1

SAFRINHA MAIZE ALONE AND COMBINED WITH COVER CROPS AFTER SOYA CULTIVATION

Roniel Giaretta
Vanderson Vieira Batista
Karine Fuschter Oligini
Lucas Link
Eduardo Júnior de Oliveira da Silva
Ana Claudia Schllemer dos Santos
Paulo Fernando Adami

Abstract: The cultivation of maize off-season has been increasing in recent years in Brazil. However, after the cultivation of maize off-season, during the off-season, it is common to find ploughed areas with unprotected soil, without the presence of cover crops, or even areas with a large number of weeds. In this context, the aim of the study was to assess the components of safrinha maize productivity and the accumulation of cover biomass after soya cultivation in different cropping systems (single and intercropped). The experiment was conducted in the municipality of Dois Vizinhos - Paraná, during the 2017 off-season. The treatments evaluated were: 1 - Maize intercropped with brachiaria in pre-sowing of maize; 2 - Maize intercropped with ryegrass sown at V8; and 3 - Maize alone. The data was compared by analysis of variance and the means by the Tukey test (0.05). Plant height, thousand grain mass and maize yield decreased in the intercropping systems, but there was an increase in dry mass accumulation in the intercropped crop. Maize plant population, number of rows per ear, number of grains per row and number of grains per ear were not influenced by the systems evaluated. Growing maize intercropped with soya reduced yields, but increased the amount of residual straw, a positive factor when considering soil protection during the off-season.

INTRODUCTION

Maize is sown in two seasons in Brazil: the summer crop, grown during the spring/summer period, and the off-season crop, grown during the autumn/winter period. According to data from CONAB (2018), combining the 2016/2017 harvest period and the off-season period, 17.5 million hectares were sown with maize in Brazil, with the off-season accounting for 68.8%, around 67.3 million tonnes of the maize produced.

The success of off-season maize productivity is directly linked to the annual result of the agricultural harvest, which involves everything from the sowing of the first crop, normally carried out with soya, combined with the cultural and nutritional management of the crop and the sowing season, which is carried out right after the harvest of the first crop. However, after the safrinha maize crop there is the off-season, the winter period, which is extremely important for the future crop, as it can interfere with the crop's productive potential and consequently the economic results.

In the off-season, it is common to find ploughed areas with unprotected soil and no cover plants, or even areas with a large number of weeds, which increase the seed bank of these weeds, increase the cost of management and jeopardise the implementation and management of the next crop. Pires et al. (2008) point out that in addition to the problems related to the increased occurrence of weeds, these areas are subject to increased soil erosion and compaction, not to mention future losses in the next harvest.

In this context, where the majority of areas remain fallow in the off-season, either due to carelessness on the part of the producer or the lack of suitable management options, studies are being carried out on cover crops in an attempt to establish a management system to solve or minimise these problems after the off-season maize crop. The use of cover crops aims to increase the amount of straw in the system during the off-season, reducing compaction and weed germination (DEBIASI et al., 2010) (PIRES et al., 2008).

However, the literature lacks studies that manage to efficiently establish cover crops associated with the off-season crop, without them harming or interfering with the yield and productivity of the off-season crop. In this sense, studies on the interaction between maize and cover crops in the system are extremely necessary, as the intercropping system can result in benefits for a continuous production system, enabling the use and sustainability of the soil, and consequently bringing benefits for the environment and the rural producer.With this in mind, the aim of this study was to assess the yield of safrinha maize after soya cultivation, grown alone and in consortium with brachiaria (Brachiaria brizantha cultivar Marandu) and ryegrass (Lollium multiflorum cultivar Barjumbo), as well as to verify the accumulation of biomass for cover resulting from these treatments.

MATERIAL AND METHODS

The experiment was conducted in the community of São Valentim, a rural area in the municipality of Dois Vizinhos, Paraná, located at latitude 25°48'19"S, longitude 53°06'28"O and at an altitude of approximately 530 metres. The region's soil is classified as Latossolo Vermelho Distrófico (BHERING, 2008). The region's climate is classified as Cfa - a humid subtropical climate without a defined dry season, with an average temperature of the warmest month of 22°C and infrequent frosts (ALVARES et al., 2013).

The rainfall (mm) and the maximum (°C) and minimum (°C) temperatures recorded during the study period are shown in Figure 1.

In order to better analyse and understand the results, soil was sampled at a depth of 0 to 10 cm before sowing the maize, and sent to the laboratory for chemical analysis, the results of which are described in Table 1. According to the producer, these good soil chemical parameters are due to the continuous use of poultry litter over the last few years in the study area, which uses an average of 5 tonnes ha^{-1} of poultry litter every two years.

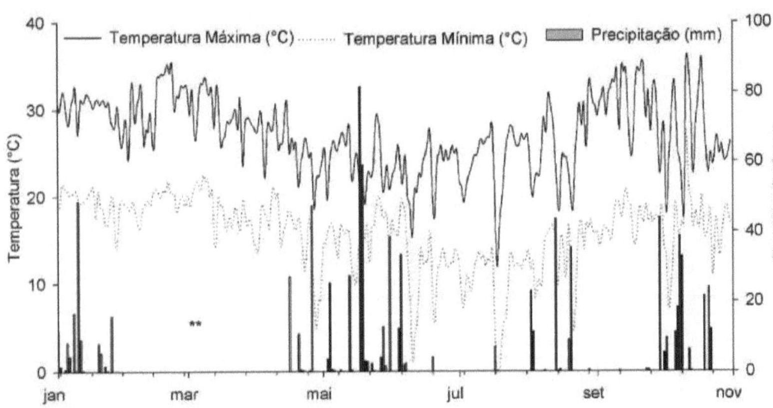

Figure 1. Maximum temperature (°C), minimum temperature (°C) and rainfall (mm) data recorded by the INMET station in Dois Vizinhos - PR, during the study period. Source: INMET (2018). **Precipitation data not recorded in February and March 2017.

Table 1. Chemical analysis of the soil in the experimental area, (0 - 10 cm), Dois Vizinhos, PR, January 2018.

M.O.	K	Ca	Mg	AL3+	P	Cu	Fe	Zn	pH	V
g/dm^3	Cmolc/dm^3 ----------------				----------mg/dm^3------------				CaCl2	(%)
52,03	1,32	8,	,	0	,5	11,45	21,86	21,13	5,4	70,16

The study was conducted in a randomised block design with four replications. The treatments evaluated were: 1 - Maize intercropped with brachiaria before sowing maize; 2 - Maize intercropped with ryegrass sown at V8; and 3 - Maize alone. The experimental units were made up of 9 rows of corn, spaced 45 cm apart and 10 metres long, totalling 40.5 m^2 .

The study used the single maize hybrid AG9030PRO3 with a super-early cycle (795 GDU) (AGROCERES, 2017), and commercial seeds of Brachiaria brizantha cultivar Marandu (Brachiaria) and Lollium multiflorum cultivar Barjumbo (ryegrass).

Safrinha maize was sown on 15 February 2017 at a density of 65,000 seeds ha^{-1} , just after harvesting the Nidera Ns5445 soya bean, which had been sown on 29 September 2016. The maize was sown using a Planti Center model 9,000, generation 3, nine rows spaced 45 centimetres apart with a conventional seed distribution system (disc + ring) attached to a New Holland TL-95E tractor. The sowing speed was 4 km h^{-1} , and solid lubricant was added to the seeds, 3 grams Kg^{-1} of powdered graphite, improving the performance and uniformity of seed deposition (CUNHA and JÚNIOR, 2012). At the same time as sowing, chemical fertiliser with the 08-20-15 formula was added at a dosage of 210 Kg ha^{-1} .

The cover crops were over-seeded manually by hand, simulating tractorised equipment so that they could be applied at the farmer's level. The brachiaria was sown in pre-planting of the maize crop, using a density of 15 kg ha^{-1} . Also, when the maize crop was at the V8 stage of development, a time when the last possible low tractor input is characterised (25 March 2017), ryegrass was sown (55 Kg ha^{-1}).

Weed control in maize was carried out at the V3 stage, so that the brachiaria was already starting to tiller. Only the herbicide Atrazine was used, where according to Jakelaitis et al. (2005), the isolated herbicide causes less phytotoxicity to the forage. A dose of 3,000 g ha^{-1} of active ingredient was used, equivalent to 6 Lt ha^{-1} , of the commercial product Atrazine Nortox 500 SC.

At stage V4 of the maize, an application of nitrogen (N) was made at a dose of 250 kg ha^{-1} of urea to supplement the plant, when favourable climatic conditions for the application were observed (MAR et al., 2003) and when the maize had grain moisture of approximately 22%, the evaluations began.

The plant population (POP) was assessed (plants ha^{-1}), obtained by counting the total number of plants present in the two central rows of the experimental units, 8 metres long, and the value was extrapolated to one hectare.

Plant height (AP) (cm) and ear insertion height (AIE) (cm) were also assessed on ten plants per experimental unit, and the arithmetic mean of the values reported was calculated and used for data analysis.

As for the yield components, the number of grains per row (NGF) and the number of rows per ear (NFE) were assessed in ten ears per experimental unit, and the average values were used to analyse the data. The number of grains per ear (NGE) was determined by multiplying the NFE and NGF in the respective experimental units.

The mass of a thousand grains (MMG) (grams) was also determined by weighing 200 grains from each experimental unit, multiplied by a correction factor of five, with a grain humidity of 13%. Productivity (PROD) (Kg ha^{-1}) was obtained by threshing the ears of corn collected from each experimental unit, consisting of 5 central rows with a length of 8 metres (18 m^2), which were then weighed, with the humidity corrected to 13%, and the productivity extrapolated to the hectare.

To assess the total accumulation of dry mass (ATMS) (Kg ha^{-1}), the material (cover crops + straw left over from the maize harvest) present in a known area of 0.25 m^2 was collected and taken to an oven to dry at 65°C. When the material had reached a constant mass, it was weighed on a precision scale and the value extrapolated to the hectare.

The data obtained was tabulated and subjected to an analysis of variance (F test) with a 5% probability level. When this was significant, the means were compared using the Tukey test at a 5% probability level, with the aid of the Sisvar 5.6 software (FERREIRA, 2008).

RESULTS AND DISCUSSIONS

Table 2 shows that the final population of maize plants was not influenced by the cultivation systems evaluated. There was an average of 56,888 plants ha^{-1} in the experimental field, i.e. there was a reduction of 12.48% between the sowing density (65,000 seeds ha^{-1}) and the final plant population. The final plant population is a key index for maintaining the productive potential of hybrids, and it is essential to establish the population within the technical recommendation in order to achieve maximum productivity.

Table 2. Plant population (POP) (plants ha^{-1}), ear insertion height (AIE) (cm), plant height (AP) (cm), number of grains per row (NGF) and number of ear rows (NFE) of the AG-9030 PRO3 maize hybrid, grown in different production systems in the 2017 off-season, in Dois Vizinhos - PR, (2018).

Treatments	POP	IEA	AP	NGF	NFE
T 1	56.888,89	91,00 b	207,67 b	27,33	16,00
T 2	56.296,29	94.67 ab	212,05 b	27,33	16,67
T 3	57.481,48	101,00 a	221,45 a	28,00	16,67
Average	56.888,89	95,56	213,78	27,56	16,44
F-value	$0,7563^{ns}$	0,0139*	0,0050**	$0,7095^{ns}$	$0,6297^{ns}$
DMS	4.751,17	7,13	8,86	2,77	2,36
CV (%)	3,33	2,98	1,52	4,01	5,73

** significant at the 1% probability level (p < .01), * significant at the 5% probability level (.01 =< p < .05) and^{ns} not significant (p >= .05). Averages with different letters in the column show significant statistical differences using the Tukey test at 5% probability. T1 - Maize intercropped with brachiaria before sowing maize; T2 - Maize intercropped with ryegrass sown at V8; and T3 - Maize alone.

When compared with other crops, such as soya and wheat, maize has little plasticity, i.e. the ability to express a second ear, while soya and wheat have the ability to express more branches and more tillers per plant, compensating for failures and/or lower populations (FORNASIERI FILHO, 2007), which is why it is essential to maintain an adequate plant stand in the field.

In this context, it is assumed that the productivity of the experimental crop may have been influenced by the plant population, but this influence occurred in all the treatments, as they showed CV (%) values of only 3.33, which demonstrates homogeneity between the treatments.

The maize-brachiaria intercropping system resulted in a lower ear insertion height (91 cm) compared to the maize grown in monoculture (101 cm) (Table 2). Table 2 also shows that plant height is lower in intercropped maize systems.

Costa et al. (2012), studying nitrogen levels and brachiaria genera in intercropping with maize in the 2008/2009 and 2009/2010 harvests, found no effects of brachiaria on plant height and ear insertion height. However, Pariz et al. (2011) observed that B. ruziziensis provided lower ear insertion height when grown in intercropping with maize. In this sense, care should be taken when setting up the maize and brachiaria consortium so that there is no competition between the plants, so that the brachiaria does not interfere with the development of the maize plant, as this could harm the crop's productivity in the future.

As for the yield components, Table 2 shows that the number of grains per row and the number of rows per ear were not influenced by the treatments evaluated, with average values of 27.56 and 16.44, respectively, in line with studies carried out by Richart et al. (2010). It was also found that the number of grains per ear was not affected by the cultivation systems evaluated (Table 3).

Table 3. Number of grains per ear (NGE), thousand grain mass (MMG) (g), productivity (PROD) (Kg ha^{-1}) and total accumulation (maize + intercropped plant) of dry mass (ATMS) (Kg ha^{-1}) of maize hybrid AG-9030 PRO 3, grown in different production systems in the 2017 off-season, in Dois Vizinhos - PR, (2018).

Treatments	NGE	MMG	PROD	ATMS
T 1	437,33	241,55 b	4.412,50 c	16.461,68 a
T 2	456,00	243,06 b	5.046,14 b	13.938,80 b
T 3	466,67	258,61 a	5.847,60 a	12.699,86 c
Average	453,33	247,74	5.102,09	14.366,89
F-value	$0,6050^{ns}$	0,0062**	0,0010**	0,0001**
DMS	87,07	11,23	599,59	1.081,01
CV (%)	7,67	1,81	4,69	3,00

** significant at the 1% probability level ($p < .01$) and^{ns} not significant ($p >= .05$). Averages with different letters in the column show significant statistical differences using the Tukey test at 5% probability. T1 - Maize intercropped with brachiaria before sowing maize; T2 - Maize intercropped with ryegrass sown at V8; and T3 - Maize alone.

However, the mass of a thousand grains decreased in the consortium systems (Table 3). It can also be seen that yields differed between the systems evaluated, with a yield of 5,847 Kg ha^{-1} , differing from the brachiaria and ryegrass intercropping systems, which had yields of 4,412.50 and 5,046.14 Kg ha^{-1} , respectively, which also differed from each other (Table 3). These results differ from those reported by Garcia et al. (2012), who emphasised in their studies that intercropping systems using maize and forage did not reduce grain yields compared to maize grown without intercropping.

The inverse statistical result to productivity is observed for the total accumulation of dry mass, in which the brachiaria consortium showed the highest accumulation, 16,461.68 Kg ha^{-1} , followed by the maize and ryegrass consortium 13,938.80 Kg ha^{-1} , with the lowest total accumulation of dry mass observed in the maize monoculture, 12,699.86 Kg ha^{-1} (Table 3).

Pariz et al. (2011) studied four species of brachiaria (Brachiaria *brizantha* cv. 'Marandu', *Brachiaria decumbens, Brachiaria ruziziensis and Brachiaria hybrid* cv. 'Mulato II') intercropped in the row and in the sowing of the corn crop and found that *Brachiaria brizantha accumulated* similar dry mass values, regardless of the sowing system, but that the other genera developed better when sown in the sowing system.

Garcia et a. (2012), on the other hand, carried out an economic analysis of the grain yield of maize, intercropped with forage plants of the *Brachiaria* and *Panicum* genera in a no-till system, and reported that the most recommended intercropping system is the use of maize with *B. ruziziensis*, with simultaneous sowing of the crops being the best form of establishment. Richart et al. (2010) also describe that the simultaneous sowing of the two species (maize + brachiaria) shows the highest dry mass yields of the cover species, contributing to these results.

As observed in this study, the intercropping system can reduce the productivity of maize crops, but it also guarantees a higher biomass accumulation. In this context, it is up to the producer to analyse the best decision to be made for his crop and choose which system, intercropping or monocropping, to use during the off-season.

CONCLUSIONS

Intercropping safrinha maize with Brachiaria brizantha cultivar Marandu (brachiaria) or Lollium multiflorum cultivar Barjumbo (ryegrass) results in lower ear insertion height, plant height, thousand grain mass and yield compared to growing maize in monoculture.

The population of maize plants, the number of grains per row, the number of rows per ear and the number of grains per ear were influenced by the systems evaluated.

Total biomass accumulation was greater with the use of consortium systems, and was higher with the use of brachiaria.

BIBLIOGRAPHICAL REFERENCES

ALVARES, C. A.; STAPE, J. L.; SENTELHAS, P. C.; MORAES, J. L. G; SPAROVEK, G.. Kõppen's climate classification map for Brazil. Meteorologische Zeitschrift, Stuttgart, v. 22, p.711-728, 2013.

AGROCERES. 2017 Safrinha seed guide. Agroceres Sementes. 2017.

BHERING, S. B., DOS SANTOS, H. G., BOGNOLA, I. A., CÚRCIO, G. R., CARVALHO JUNIOR, W. D., CHAGAS, C. D. S., SILVA, J. D. S. Soil map of the State of Paraná, updated legend. In: Embrapa Florestas-Article in conference proceedings (ALICE). In: BRAZILIAN CONGRESS OF SOIL SCIENCE, 32nd, 2009, Fortaleza. Soil and bioenergy production: perspectives and challenges: proceedings [Viçosa, MG]: SBCS; Fortaleza: UFC, 2009., 2008.

CONAB. National Supply Company. Monitoring the 2017/18 Brazilian grain harvest - 4th Conab survey. Brasília - DF. p. 132, v. 5; January 2018.

COSTA, N, R.; ANDREOTTI, M.; GAMEIRO, R. A.; PARIZ, C. M.; BUZETTI, S.; LOPES, K. S. M. Nitrogen fertilisation in the intercropping of maize with two brachiaria species in a no-till system. Pesquisa Agropecuária Brasileira, Brasília, v.47, n.8, p.1038-1047, Aug. 2012.

CUNHA, J. P. A. R.; JÚNIOR, R. L. S. Congresso Nacional de Milho e Sorgo, XXIX. Performance of a no-till seeder in maize cultivation, Águas de Lindóia, p. 3227 - 3232, 2012.

DEBIASI, H., LEVIN, R., CONTE, O., KAMIMURA, K. M. Soya and maize productivity after winter cover crops and mechanical soil decompaction, Pesquisa Agropecuária Brasileira, Brasília, v.45, n.6, p. 603-612, 2010.

FERREIRA, D. F. Sisvar, a programme for statistical analysis and teaching. Revista Científica Symposium, 6:36-41, 2008.

FORNASIERI FILHO, D. Manual da cultura do milho. Jaboticabal-SP, Funep, p. 576, 2007.

GARCIA, C. M. P.; ANDREOTTI, M. TARSITANO, M. A. A.; TEIXEIRA FILHO, M. C. M.; LIMA, A. E. S.; BUZETTI, S. Economic analysis of the grain yield of maize intercropped with forage plants of the genera Brachiaria and Panicum in a no-till system. Revista Ceres, Viçosa, v. 59, n.2, p. 157-163, Mar/Apr, 2012.

INMET. National Meteorological Institute. Available at: <http://INMEt.gov.br/sonabra/pg_dspDadosCodigo_sim.php?QTg0Mw==>. Accessed on: 07 January 2018.

JAKELAITIS, A.; SILVA, A. F.; SILVA, A. A.; FERREIRA, L. R., FREITAS, F. C. L.; VIVIAN, R.; Influence of Herbicides and Sowing Systems of *Brachiaria brizantha*

Consortium with Maize, Revista Planta Daninha, Viçosa - MG, v.23, n.1, p. 59-67, 2005.

MAR, G. D.; MARCHETTI, M. E.; SOUZA, L. C. F.; GONÇALVES, M. C.; NOVELINO, J. O. Part of the first author's master's thesis in Agronomy/Plant Production. Production of Safrinha Maize as a Function of Doses and Times of Nitrogen Application, Bragantina, Campinas, v.62, n.2, p. 267 - 274, 2003.

PARIZ, Cristiano M.; ANDREOTTI, Marcelo; AZENHA, Mariana V.; BERGAMASCHINE, Antonio F.; MELLO, Luiz M. M.; LIMA, Ronaldo C. Maize grain yield and dry mass of brachiaria in consortium in the crop-livestock integration system. Ciência Rural, v.41, n.5, mai, 2011.

PIRES, Fábio R.; ASSIS, Renato L.; PROCÓPIO, Sérgio O.; SILVA, Gilson P.; MORAES, Leonardo L.; RUDOVALHO, Marcos C.; BÔER, Carlo A.; Revista Ceres. Agronomic characteristics of maize cultivars as a function of plant populations and sowing times, Viçosa - MG, v.55, n.2, p. 94 - 101, 2008.

RICHART, A.; PASLAUSKI, T.; NOZAKI, M. H.; RODRIGUES, C. M.; FEY, R. Performance of safrinha maize and Brachiaria ruziziensis cv. Comum in consortium. Revista Brasileira de Ciências Agrárias. Recife - PE, v.5, n.4, p. 497 - 502, 2010.

CHAPTER 2

OFF-SEASON COVER CROPS AND THEIR POTENTIAL FOR DRY MASS ACCUMULATION

Roniel Giaretta
Karine Fuschter Oligini
Vanderson Vieira Batista
Lucas Link
Ana Claudia Schllemer dos Santos
Cleverson Alessio da Silva
Paulo Fernando Adami

Abstract: Mulch is of great importance to the no-till system and also serves as protection for the soil during the off-season. The aim of this study was to assess the dry mass accumulation potential of different cover species in different cropping systems. The study was conducted in the municipality of Dois Vizinhos, Paraná, in a randomised block design. Twelve cropping systems were evaluated, some after beans and others after soya. After bean cultivation in the harvest, maize in the off-season and sowing of white oats, black oats, turnips and fallow, also brachiaria over-seeding in pre-harvest of beans (intercropped with maize), brachiaria in pre-sowing of maize and brachiaria over-seeding in pre-harvest of beans (single). After soya cultivation in the harvest period, we evaluated the cultivation of intercropped maize, white oats and ryegrass in V8 of the maize, fallow area, brachiaria in pre-sowing of maize and brachiaria in pre-sowing of maize (single). The dry mass accumulation of the cover species and the total dry mass accumulation (cover species + maize straw) were evaluated. After growing beans in the harvest, maize in the off-season and black oats, white oats, brachiaria in over-seeding and brachiaria in pre-sowing maize, there was a greater accumulation of dry mass. After growing soya beans, maize intercropped with brachiaria before planting showed higher dry mass production. Growing brachiaria during the off-season provides a greater accumulation of dry mass among the species and cover, but if corn is grown in the off-season, there is a greater accumulation of total dry mass in these cropping systems. Among all the systems studied, growing beans in the harvest period, with maize in the off-season and sowing white oats in the winter period, results in greater dry mass accumulation when sowing the next crop.

INTRODUCTION

The cultivation of off-season maize in the south-western region of Paraná has been gaining ground, with many producers opting to plant soya in the harvest, followed by off-season maize. However, during the off-season, i.e. until the new soya harvest, they leave the area fallow. This period in which the soil is left idle for consecutive years can result in problems that are often difficult to overcome. According to Gomes (2016) and Rossetti and Centurion (2015), these areas can result in lower productivity and higher production costs, due to the need for management with herbicides, fertilisers and mechanical intervention

In this context, the use of cover crops in consortium with the off-season crop enters the agricultural system for various purposes, employing conservationist soil management to maintain the system (WALTRICK, 2010). Also, according to the researcher, the straw avoids problems with erosion, since the consortium provides an increase in straw in the management of the system, which remains on the area after the crop of interest, prolonging soil cover, considerably reducing erosion rates and reducing losses due to climatic abnormalities in subsequent years.

For Debiasi et al. (2010), when considering the productivity of soya and maize in the following year in no-till areas, they found that the use of cover crops during the winter period is a more viable method for circumventing the effects of soil compaction, compared to mechanical scarification and sowing at a greater depth with the furrowing tines. Therefore, in addition to reducing the compaction index, the greater coverage with plant residues also acts to maintain water on the soil surface, providing greater availability time, since the system has reduced evaporation rates with protected soil (ANDRADE, 2008).

Increasing and/or maintaining straw on the soil in a no-till system also provides results in weed control. This practice is associated with integrated weed management (IPM), which aims to manage crops within crop rotation, seeking to increase dry mass, which provides greater soil cover, making it difficult for weeds to germinate that may develop in the off-season (GOMES and CHRISTOFFOLETI, 2008).

However, the production of dry mass from cover crops varies according to the species and the management adopted in the field. The most efficient weed suppression/control is achieved with higher levels of residual mulch, i.e. plants that produce a greater amount of dry mass per hectare provide greater inhibition of weed germination (CORREIA et al., 2006). Black oats, ryegrass and turnip rape are winter cover crops that have a high dry mass and root content, and are important species in terms of nutrient accumulation in their aerial part, providing nutrition for the successor crop, along with soil protection, lowering soil temperature and retaining more moisture in the soil (WOLSCHICK et al., 2016).

The method of intercropping maize with brachiaria also stands out. Based on the Santa Fe System, this system has been widely used and studied with a view to improving cultivation methods (KLUTHCOUSKI et al., 2000). The purpose of this system is to produce maize grains and fodder for animal feed, and brachiaria can also be used as a mulching option, as it provides a large accumulation of dry mass, ideal for soils with low levels of straw (RICHART et al., 2010).

Knowing the importance of mulch for the no-till production system and the problems that areas without mulch can present during the off-season, the aim of this study was to assess the dry mass accumulation potential of different mulch species in different cropping systems.

MATERIAL AND METHODS

The study was carried out in a ploughed area located in the community of São Valentim, municipality of Dois Vizinhos, Paraná, latitude 25°48'19"S, longitude 53°06'28"O. The area has an average altitude of approximately 530 metres and is classified as Latossolo Vermelho Distrófico (BHERING, 2007), with a Cfa climate (ALVARES et al., 2013) and an average annual rainfall of 2,044 mm (POSSENTI et al., 2007).

The study was conducted in a randomised block design with three replications, and 12 treatments were evaluated. The treatments consisted of different production systems in the harvest and off-season periods, and the dry mass accumulation of different cover species was evaluated in the off-season. Table 1 shows the treatments evaluated, as well as the crop grown during the harvest and off-season and the cover crop species grown during the off-season.

Table 1. Cropping systems (treatments) evaluated, considering crops from the harvest period, off-season crops and cover species in the off-season, Dois Vizinhos (2018).

Treatment	Safra	Harvest	Off-season
T 1	Beans	Maize$^{(1)}$	White oats$^{(6)}$

T 2	Beans	Maize[1]	Black oats[6]
T 3	Beans	Maize[1]	Turnip[6]
T 4	Beans	Maize[1]	Fallow land
T 5	Beans	Maize + Brachiaria[1][3]	Brachiaria
T 6	Beans	Maize + Brachiaria[1][4]	Brachiaria
T 7	Beans	Brachiaria[3]	Brachiaria
T 8	Soya	Maize[2]	White oats[5]
T 9	Soya	Maize[2]	Ryegrass[5]
T 10	Soya	Maize[2]	Fallow land
T 11	Soya	Maize + Brachiaria[2][4]	Brachiaria
T 12	Soya	Brachiaria[4]	Brachiaria

(1) Sowing maize after bean harvest (03/01/2017). (2) Sowing maize after soya harvest (15/02/2017). (3) Pre-harvest brachiaria over-sowing of beans (27/12/2016). (4) Brachiaria in pre-sowing of maize (03/01/2017 - T 6) (15/02/2017 - T 11 and T 12). (5) Over-sowing of cover crops when the maize was at V8. (6) Sowing of cover crops.

As for the cover species, commercial seeds were used and planted at densities of 20 kg ha^{-1} for brachiaria (Brachiaria brizantha

cultivar Marandu), 75 Kg ha^{-1} of black oats (Avena strigosa cultivar BRS 139), 50 Kg ha^{-1} for ryegrass (lollium multiflorum cultivar Barjumbo), 80 of white oats (Avena sativa cultivar URS Taura) and 20 Kg ha^{-1} for common turnip rape (Raphanus sativus).

The cover crops were sown after the harvest of the maize sown on the bean stubble (the first season of safrinha maize) using a SHM15/17 (Semeato) small grain seeder, comprising 17 rows spaced 19 centimetres apart. The cover crops for the other treatments were sown manually and by hand.

On 13 October 2017, the dry mass accumulation of the cover crops was assessed, as well as the remaining mass of the maize plants resulting from the off-season crop. The dry mass assessments were carried out by collecting the material present in a known area of 0.25 m^2 , subsequently drying this material in an oven at 65°C, then weighing the mass on a precision scale and extrapolating the value to hectare (Kg ha^{-1}).

The data analysis considered the dry mass accumulation values of the cover crops and the total dry mass accumulation (dry mass of cover crops + dry mass of safrinha corn). The analysis was carried out considering the post-bean (T1 to T7) and post-soya (T8 to T12) cropping systems. Statistics were also analysed for all treatments (T1 to T12). The data obtained was tabulated and subjected to analysis of variance (F test) with a 5% probability level and when this was significant, the means were compared using the Skott-Knott test at a 5% probability level, with the aid of Assistat 7.7beta software (SILVA e AZEVEDO, 2002).

RESULTS AND DISCUSSIONS

Table 2 shows the treatments on the bean-growing area during the harvest period. It can be seen that the treatment consisting only of brachiaria (T 7) during the off-season showed a higher accumulation of dry mass of the cover crop (7,961 Kg ha^{-1}) than the other treatments. Next, with cover crop DM accumulation of 6,088, 4,812, 5,639 and 5,810 Kg ha^{-1} , were the treatments black oats (T 1) (Figure 1 B), white oats (T 2) (Figure 1 A), brachiaria in bean over-sowing (T 5) and brachiaria in maize pre-sowing (T 6). On the other hand, the lowest DM accumulation of the cover species was observed with the use of turnip (T 3) (3,402 Kg ha^{-1}) (Figure 1 C), followed by the fallow treatment after safrinha maize cultivation (T 4), in which the weeds that appeared in the system presented 351 Kg ha^{-1} of dry mass (Table 2).

Table 2. Dry mass accumulation of different cover species (DM Cover) (Kg ha^{-1}) and total dry mass accumulation (DM Total) (Kg ha^{-1}) (cover species + maize straw), in different off-season maize production systems, after beans in the harvest period. Dois Vizinhos - PR (2018).

Treatment	MS coverage	MS Total
T 1	6.088 b	19.260 a
T 2	4.812 b	17.984 a
T 3	3.402 c	16.573 b
T 4	351 d	13.172 c
T 5	5.639 b	18.257 a
T 6	5.810 b	18.571 a
T 7	7.961 a	7.961 d
Average	4.866	15.968
F-value	59,11**	126,16**
CV (%)	11,83	4,93

** significant at the 1% probability level ($p < .01$). Averages with different letters in the column show significant statistical differences using the Scott-Knott test at 5% probability.

Although T 7, brachiaria cover without off-season crops, had the highest dry mass accumulation among the cover plant species, this treatment had the lowest total dry mass accumulation (7,961 Kg ha^{-1}), even lower than the fallow treatment, which had total DM values of 13,172 Kg ha^{-1} (off-season maize straw + weeds) (Table 2).

The highest accumulation of total DM was observed in the treatments with black oats (T 1), white oats (T 2), brachiaria in bean over-sowing (T 5) and brachiaria in maize pre-sowing (T 6) (Table 2). Next, with a total DM accumulation of 16,573 Kg ha^{-1} , the treatment consisting of turnip (T 3) stands out in systems with bean crops in the harvest period (Table 2).

Figure 1 - Experimental areas during the winter period, which contained maize intercropped with white oats (A), black oats (B) and turnips (C). Source: Author (2018).

In season 2, after soya in the harvest period, all the treatments analysed showed statistical differences between them in terms of the accumulation of DM from the cover crops (Table 3). With the highest DM accumulation, T 12 stood out, composed of brachiaria in monoculture (7,491 Kg ha^{-1}), followed by maize + brachiaria in pre-planting (4,275 Kg ha^{-1}). With a production of 2,439 Kg ha^{-1} of DM, the treatment composed of white oats stands out, while the treatment using ryegrass as a cover crop produced 1,906 Kg ha^{-1} of DM (Table 3). The treatment that remained fallow (T 10) during the off-season produced 255 Kg ha^{-1} of DM, which came from weeds that appeared in the area (Table 3).

Table 3 shows that the corn / brachiaria pre-planting treatment (16,462 Kg ha^{-1}) had the highest accumulation of Total DM among the post-soya treatments. Maize / white oats and maize / ryegrass, with 15,138 and 14,606 Kg ha^{-1} Total DM respectively, did not differ from each other, but were superior to using maize in monoculture, which accumulated 12,700 Kg ha^{-1} of Total DM (Table 3). The lowest accumulation of Total DM after soya was observed in the treatment using only brachiaria (7,491 Kg ha^{-1}) (Table 3).

Table 3. Dry mass accumulation of different cover species (DM Cover) (Kg ha^{-1}) and total dry mass accumulation (DM Total) (Kg ha^{-1}) (cover species + maize straw), in different off-season maize production systems, after soya in the harvest period. Dois Vizinhos - PR (2018).

Treatment	MS Coverage	MS Total
T 8	2.439 c	15.138 b
T 9	1.906 d	14.606 b
T 10	255 e	12.700 c
T 11	4.275 b	16.462 a
T 12	7.491 a	7.491 d
Average	3.273	13.279
F-value	398,32**	399,68**
CV (%)	7,63	5,29

** significant at the 1% probability level (p < .01). Averages with different letters in the column show significant statistical differences using the Scott-Knott test at 5% probability.

When we statistically compare the DM accumulation of the cover crops from the two seasons together, we realise that the analysis divides the treatments into six different groups (Table 4). The first group (A), with the highest DM accumulation, includes the treatments composed of brachiaria in monoculture after beans (7,961 Kg ha^{-1}) and after soya (7,461 Kg ha^{-1}) (Table 4).

The second group (B) was made up of the treatments maize with brachiaria in pre-planting, maize with brachiaria in over-sowing and maize with white oats in monoculture, both treatments being post-bean, with an accumulation of DM of 5,339 Kg ha^{-1} , 5,810 Kg ha^{-1} and 6,088 Kg ha^{-1} respectively (Table 4). With 4,812 Kg ha^{-1} and 4,275 Kg ha^{-1} of DM, the third group (C) was made up of the maize / black oats in season 1 and maize / brachiaria (pre-planting) treatments respectively and the fourth group (D) is represented by the treatment using turnip in season 1, which accumulated 3,402 Kg ha^{-1} (Table 4). The fifth group (E) is made up of the treatments using white oats (2,439 Kg ha^{-1}) and ryegrass (1,906 Kg ha^{-1}) as cover in season 2, and the sixth group (F) is made up of the treatments without cover crops, fallow, which showed DM accumulation of 351 and 255 Kg ha^{-1} after beans and after soya respectively (Table 4).

Table 4. Dry mass accumulation of different cover species (DM Cover) (Kg ha^{-1}) and total dry mass accumulation (DM Total) (Kg ha^{-1}) (cover species + maize straw), in different off-season maize production systems. Dois Vizinhos - PR (2018).

Treatments	MS coverage	MS Total
T 1	6.088 B	19.260 A
T 2	4.812 C	17.984 B
T 3	3.402 D	16.573 C
T 4	351 F	13.172 E
T 5	5.639 B	18.257 B
T 6	5.810 B	18.571 B
T 7	7.961 A	7.961 F
T 8	2.439 E	15.138 D
T 9	1.906 E	14.606 D
T 10	255 F	12.700 E
T 11	4.275 C	16.462 C
T 12	7.491 A	7.491 F
Average	4.152	14.848
F-value	97,50**	169,44**
CV (%)	11,24	4,52

** significant at the 1% probability level ($p < .01$). Averages with different letters in the column show significant statistical differences using the Scott-Knott test at 5% probability.

When we statistically analyse the two seasons together, we can see in Table 4 that the highest accumulation of Total DM was recorded in the maize/white oat treatment after beans (19,260 Kg ha^{-1}). The group made up of maize/black oats, maize/brachiaria pre-planting and maize/brachiaria at desiccation, both after beans, appear in the second group (B), with Total DM accumulation of 17,984, 18,257 and 18,571 Kg ha^{-1} , respectively (Table 4). The post-bean maize / turnip (16,573 Kg ha^{-1}) and pre-plant maize / brachiaria (16,462 Kg ha^{-1}) treatments make up the third group (C) (Table 4). The fourth group (D) is made up of the treatments maize / white oats (15,138 Kg ha^{-1}) and maize / ryegrass (14,606 Kg ha^{-1}), both planted after soya (Table 4). The treatments in which maize was grown in monoculture formed the fifth group (E), with a total DM accumulation of 13,172 Kg ha^{-1} after beans and 12,700 Kg ha^{-1} after soya (Table 4). The treatments that had the soil occupied only by brachiaria had the lowest accumulation of Total DM and formed the sixth group (E), with an accumulation of 7,961 Kg ha^{-1} of Total DM after beans and 7,491 Kg ha^{-1} after soya (Table 4).

Agostini (2016), evaluating the accumulation of dry mass (DM) of different forage species grown in consortium with maize in the off-season, observed that the treatment composed of P. Maximum cv. Aruanateve obtained a lower yield of DM after the maize harvest compared to the treatment using B. Ruziziensis and B. Aruanateve. Ruziziensis and B. Brizanta, but the total accumulation (maize + cover crop) was higher than the control treatment. The researcher also found that the Maize+B.Brizantha cv. Piatã consortium shows good results both in terms of forage DM production and maize grain yield.

Wolschick et al. (2016) evaluated various cover species during the winter period and concluded that soil cover is influenced by the type of plant species and growth period, in which at the beginning of the cycle it is greater when fodder turnip and a consortium of species (oats, vetch and turnip) are used, but at the end it is greater with common vetch and the consortium. The researchers also observed that when growing black oats, common vetch and the consortium, there is a greater production of dry mass of the aerial part, while black oats produce a greater mass of roots.

It can also be said that soil cover in the winter period is of fundamental importance, as it provides a series of benefits for the soil, for the production system and consequently for the farmer, such as: reduction of weeds (PIRES et al., 2008), (CORREIA et al., 2006), (GOMES and CHRISTOFFOLETI, 2008), 2008), (CORREIA et al., 2006), (GOMES and CHRISTOFFOLETI, 2008), lower risk of erosion (PIRES et al., 2008) (DEBIASI et al., 2010), soil decompression (ANDRADE, 2008), (PIRES et al., 2008), (DEBIASI et al., 2010), increased productivity (DEBIASI et al., 2010), maintenance of water on the soil surface (ANDRADE, 2008), (WOLSCHICK et al., 2016), accumulation of nutrients (WOLSCHICK et al., 2016), as well as being used in animal feed (RICHART et al., 2010), depending on the species used, In this study, it was found that the consortium systems provided an accumulation of biomass (straw) for the following harvest (Figure 2 A), much greater than the monoculture maize system (Figure 2 B), which had a large number of weeds (Figure 2 C), which needed to be controlled for the next harvest.

Figure 2: Sowing of the crop after the study was carried out on the safrinha consortium area (A), safrinha maize area in monoculture (B), and weeds in the safrinha maize system in monoculture (C). Source: Author (2018).

As the dry mass production of cover crops differs, and because some species have peculiar characteristics, such as soil decompaction, biological nitrogen fixation and use in animal feed, it is up to the farmer to analyse and make a decision about which species to use, opting for the one that best suits his production system. The plants also have the potential to protect the soil, reducing erosion and the number of weeds in the crop, as well as providing organic matter for the system. However, further studies should be carried out to help identify which species to use, when to plant them and how to manage these associated plants, so that there is no competition for nutrients, water, light, etc. in the system.

CONCLUSIONS

The use of bean crops after the harvest, with maize in the off-season and black oats, white oats, brachiaria over-seeding the beans and brachiaria pre-seeding the maize, resulted in greater dry mass accumulation. On the other hand, after growing soya beans in the harvest period, and corn intercropped with pre-planted brachiaria, there was a greater accumulation of dry mass.

Brachiaria cultivation during the off-season provides a greater accumulation of dry mass between species and cover, whether on bean or soya areas during the harvest period.

However, if maize is grown in the off-season, either in a consortium or sown with species such as black oats, white oats and turnips, there is a higher accumulation of total dry mass in these cropping systems.

Of all the systems studied, growing beans in the harvest period with maize in the off-season and sowing white oats in the winter period results in a greater accumulation of dry mass when sowing the next crop.

BIBLIOGRAPHICAL REFERENCES

AGOSTINI, A. C. Master's thesis in Zootechnics. Performance of maize in consortium with different tropical forage species. Dois Vizinhos - PR, 2016.

ALVARES, C. A.; STAPE, J. L.; SENTELHAS, P. C.; MORAES, J. L. G; SPAROVEK, G. Kõppen's climate classification map for Brazil. Meteorologische Zeitschrift, Stuttgart, v. 22, p.711-728, 2013.

ANDRADE, J. G. Master's thesis in Soil Science. Water loss by evaporation from a soil cultivated with maize in no-till and conventional tillage systems, Santa Maria - RS, p.93, 2008.

BHERING, S. B., SANTOS, H. G., BOGNOLA, I. A., CÚRCIO, G.R., CARVALHO JUNIOR, W. D., CHAGAS, C. D. S., SILVA, J. D. S. In: Embrapa Florestas-

Article in conference proceedings (ALICE). In: BRAZILIAN CONGRESS OF SOIL SCIENCE, 32nd, 2009, Fortaleza. Soil and bioenergy production: perspectives and challenges: proceedings [Viçosa, MG]: SBCS; Fortaleza: UFC, 2009., 2008.

CORREIA, N. M.; DURIGAM, J. C.; KLINK, U. P. Influence of the type and quantity of plant residues on weed emergence. Weeds. Viçosa - MG; v.24, n.2, p.245-253, 2006.

DEBIASI, H., LEVIN, R., CONTE, O., KAMIMURA, K. M. Soya and maize productivity after winter cover crops and mechanical soil decompaction. Pesquisa Agropecuária Brasileira, Brasília, v.45, n.6, p. 603-612, 2010.

GOMES, L. J. P. Master's thesis in Agronomy. Response variability of different *Digitaria insularis* populations and management strategies for a resistant population, Jaboticabal - SP, p. 61, 2016.

GOMES, F. G.; CHRISTOFFOLETI, P. J. Biology and management of weeds in no-till areas.

Planta Daninha Magazine. Viçosa - MG, v.26, n.4, p. 789798, 2008.

KLUTHCOUSKI, J.; COBUCCI, T.; AIDAR, H.; YOKOYAMA, L. P.; OLIVEIRA, I. P.; COSTA, J. L. S.; SILVA, J. G.; VILELA, L.; BARCELLOS, A. O.; MAGNABOSCO, C. U.; Technical Circular Embrapa Arroz e Feijão Sistema Santa Fé - Embrapa technology: integration of crop and livestock farming through the intercropping of annual crops with forage, in ploughed areas, in the no-till and conventional systems, Santo Antônio de Goiás - GO, n. 38, p. 28, 2000.

PIRES, F. R.; ASSIS, R. L.; PROCÓPIO, S. O.; SILVA, G. P.; MORAES, L. L.; RUDOVALHO, M. C.; BÔER, C. A. Agronomic characteristics of maize cultivars as a function of plant populations and sowing times, Revista Ceres. Viçosa - MG, v.55, n.2, p. 94 - 101, 2008.

POSSENTI, J. C.; GOUVEA, A.; MARTIN, T. N.; CADORE, D. Distribution of rainfall in Dois Vizinhos, Paraná, Brazil. *In:* Seminar on Agricultural Production Systems, 1Dois Vizinhos, October 2007.

ROSSETTI, K.V.; CENTURION, J. F. Test of compaction in latosol cultivated with maize under different periods of adoption of types of management, Revista Brasileira de Ciências Agrarias. Recife - PE, v. 10, n.4, p. 499 - 505, 2015.

RICHART, A.; PASLAUSKI, T.; NOZAKI, M. H.; RODRIGUES, C. M.; FEY, R. Performance of safrinha maize and *Brachiaria ruziziensis* cv. Comum in consortium, Revista Brasileira de Ciências Agrárias. Recife - PE, v.5, n.4, p. 497 - 502, 2010.

SILVA, F. A. S.; AZEVEDO, C. A. V. de. Version of the Assistat computer programme for the Windows operating system. Revista Brasileira de Produtos Agroindustriais, Campina Grande, v.4, n.1, p.71-78, 2002.

WALTRICK, P. C.; Postgraduate dissertation in Soil Science. Rainfall erosivity in Paraná: Update, Influence of "EI Ninõ" and "La Ninã" and estimation for future climate scenarios, Curitiba, p. 106, 2010.

WOLSCHICK, N. H.; BARBOSA, F. T.; BERTOL, I.; SANTOS, K. F.; BAGIO, B.; Soil cover, biomass production and nutrient accumulation by cover plants, Revista de Ciências Agroveterinárias, Lages - SC, v.15, n.2, p. 134143, 2016.

CHAPTER 3

Brachiaria sp. DORM COVER: LITERATURE REVIEW

Vanderson Vieira Batista
Cleverson Alessio da Silva
Ana Claudia Schllemer dos Santos
Karine Fuschter Oligini
Eduardo Júnior de Oliveira da Silva
Lucas Link
Paulo Fernando Adami

Abstract: Species of the genus Brachiaria sp. are increasingly cultivated in Brazil. However, there has also been a large number of degraded pastures in the country, and the dormancy of Brachiaria seeds, which can result in low plant emergence in the field, is one of the factors responsible for this problem. Based on this context, the aim of this study is to survey the literature on the methods used to reduce the effect of dormancy in Brachiaria sp. seeds, and to check whether these methods are really efficient. Among the dormancy-breaking methods found in the literature, the following stood out: scarification with sulphuric acid (H_2SO_4), use of potassium nitrate (KNO_3) and gibberellic acid (GA_3), boiling water, storage time and temperature on germination. It can be seen that all the methods had some positive effect on reducing the dormancy of Brachiaria seeds, but the opposite effects were also observed, as well as problems in the development of the plants in sequence. In addition, there are some differences between some studies, which indicates the possibility of more than one dormancy mechanism acting on the seeds.

INTRODUCTION

The use of species of the genus Brachiaria sp. has increased in recent years (CARDOSO et al., 2014), and they stand out as one of the main forage crops in integrated crop-livestock systems, as they have a high biomass accumulation (GARCIA et al., 2013) (PETTER et al., 2013), 2013), and can be grown in consortium with maize (SCHEIDT et al., 2014) (GARCIA et al., 2013), as a soil mulch (PETTER et al., 2013), or as pasture for animal feed (FABRICE et al. 2015) (CARDOSO et al., 2014).

According to Cardoso et al. (2014), the expansion of pasture areas cultivated with species of the Brachiaria genus has increased considerably in Brazil due to their hardiness, which allows them to be cultivated in the most varied climate and soil conditions. However, around 80 per cent of Brazil's pasture areas are in some form of degradation, and the low germination rate of forage species is one of the causes of these problems (LACERDA et al., 2010).

For Lacerda et al. (2010), it is known that there are Brachiaria sp. seeds that do not germinate even when placed in favourable germination conditions, and these are classified as dormant. Costa et al. (2011) point out that most tropical forage grasses are affected by seed dormancy, which makes it difficult to determine the physiological quality of seed batches and can compromise seedling emergence and the establishment of pastures in the field. In addition, results obtained by Dias and Alves (2008) show that there can be variations in the intensity of dormancy within the same batch of seeds, which is due to the large number of seeds.

Binotti et al. (2014) commented that there may be inhibitors in the outer layers of Brachiaria sp. seeds, which can release substances or prevent water from entering, making germination difficult. The researchers also point out that soaking the seeds in certain solutions for a certain period can help to make the envelope more permeable, thus speeding up the germination process in species that have a long germination period, such as Brachiaria.

The Rules for Seed Analysis (RAS) recommend scarifying Brachiaria sp. seeds with sulphuric acid (H_2SO_4) as a way of overcoming dormancy. There is also an indication to use an aqueous solution containing potassium nitrate (KNO3 - 0.2%) as a substrate in the germination test (BRASIL, 2009).

According to Binotti et al. (2014), the dormancy of Brachiaria seeds and the difficulty in overcoming it may be related to different dormancy mechanisms acting on the seed. For these researchers, studies aimed at increasing the germination percentage and reducing the time between sowing and seedling emergence are extremely important.

Knowing that the plantability of Brachiaria sp., whether in a consortium system or in isolated systems, can be affected by factors related to seed dormancy, several studies have been carried out to identify ways and methods of reducing the percentage of dormant seeds in Brachiaria sp. Based on this context, the aim of this study is to survey the literature on what has been used to reduce the effect of dormancy in Brachiaria sp. seeds, as well as to check whether the methods used are really efficient.

DEVELOPMENT

Among the most studied methods are scarification with sulphuric acid (H_2SO_4) (LACERDA et al., 2010), (DIAS and ALVES, (COSTA et al., 2011), (ZUFFO et al., 2014), (CARDOSO et al., 2014), (CHIODINI and CRUZ-SILVA, 2013), (SILVA et al., 2014), (BINOTTI et al., 2014), 2014), (BINOTTI et al., 2014); the use of potassium nitrate (KNO_3) (COSTA et al., 2011) (BINOTTI et al., 2014), (BATISTA et al., 2015), and gibberellic acid (GA_3) (LACERDA et al., 2010), (SILVA et al., 2013), (BATISTA et al. (2015). Also noteworthy are studies involving the use of boiling water (FOWLER and BIANCHETTI, 2000), (LACERDA et al., 2010), storage time (COSTA et al., 2011), (DIAS and ALVES, 2008) (ALVES et al., 2017), (CARDOSO et al., 2014) and temperature on germination (CHIODINI and CRUZ-SILVA, 2013).

Gibberellic acid (GA_3)

In studies carried out by Lacerda et al. (2010), evaluating the use of gibberellic acid (100 and 200 mg L^{-1} - for 48 hours) on the germination of Brachiaria brizantha cv. Marandu seeds, no significant effect was found.

However, Silva et al. (2013) evaluated gibberellin concentrations (0 to 100 mg L^{-1}) and found maximum germination of *Brachiaria brizantha* 'MG 5' and 'Marandu' seeds, using doses of 57 and 62 mg L^{-1}, respectively.

Based on these studies, it can be seen that the use of gibberellic acid can have different effects on the germination of *Brachiaria* sp. seeds, and that sometimes the use of higher concentrations is detrimental to the germination of the seeds of this species. These results may be related to the fact that Brachiaria has different forms of dormancy, and knowing that GA3 is essentially aimed at overcoming the physiological dormancy of seeds, the hormone may not always have the potential to have a beneficial effect on breaking dormancy and germinating *Brachiaria* seeds.

Sulphuric acid (H_2SO_4)

Scarifying seeds using sulphuric acid is a technique widely used to break dormancy in various species. The Seed Analysis Rule (RAS) recommends scarifying *Brachiaria* sp. seeds with concentrated sulphuric acid (H_2SO_4) for a maximum of 15 minutes as a way of removing seed dormancy (BRASIL, 2009).

There is a vast amount of literature evaluating the use of H_2SO_4 on the breaking of dormancy and germination of *Brachiaria* sp. seeds. Also, some studies point to the negative effect of sulphuric acid on the initial development of *Brachiaria* sp. plants.

According to Lacerda et al. (2010), the use of concentrated sulphuric acid for five and ten minutes on *Brachiaria brizantha* cv. Marandu seeds is not efficient in breaking dormancy, with germination percentage and germination speed index being similar to the control treatment.

Dias and Alves (2008) evaluated 15-minute sulphuric acid scarification of Brachiaria brizantha cv. Marandú and found that the acid can improve or worsen the germination of forage seeds, depending on the physiological state of the seed batch. The researchers noted that the same batch can contain seeds with different degrees of dormancy, so although sulphuric acid has favoured overcoming dormancy in some situations, it is not effective in completely overcoming seed dormancy.

Similar results were observed by Costa et al. (2011), who evaluated Brachiaria humidicola seeds at 9 and 21 months of storage. The researchers found dormancy percentages of 84 and 36 per cent for seeds without scarification and with sulphuric acid scarification, respectively, at 9 months of storage, and at 21 months of storage, no significant differences were reported for the number of dormant seeds. This result shows that sulphuric acid can have different effects on breaking the dormancy of Brachiaria sp. seeds.

According to Zuffo et al. (2014), the use of scarification is an alternative for reducing dormancy in Brachiaria dictyoneura cv. Llanero seeds. According to Cardoso et al. (2014), Brachiaria brizantha cv. MG-5 seeds subjected to chemical scarification with sulphuric acid overcome dormancy, but are more susceptible to deterioration.

Chiodini and Cruz-Silva (2013), on the other hand, reported that Brachiaria brizantha cv. Marandu (Hochst. ex A. Rich.) Stapf seeds submitted to immersion in sulphuric acid showed higher germination and germination speed index than untreated seeds. Silva et al. (2014) also found that the use of sulphuric acid (98.0%) with the seeds immersed for 15 minutes, washed and dried, provided a higher germination rate of Brachiaria brizantha (Hochst. ex A. Rich) Stapf, compared to the use of hot water (60 °C) and potassium nitrate.

Binotti et al. (2014) also reported that chemical scarification is a viable technique for overcoming the dormancy of Brachiaria brizantha cv. MG-5 seeds, as it was observed that the chemical scarification treatment, which contained
69% of viable seeds, 68% germinated, i.e. 99% of viable seeds germinated. As for the treatment without scarification, of the 64% of viable seeds, only 54% germinated, i.e. germination of only 84% of the viable seeds present in the germination test.

Despite the positive effects of chemical scarification, Binotti et al. (2014) found that the procedure negatively interferes with initial seedling growth. According to the researchers, the scarified seeds may have allocated energy from the reserve tissue to reorganising and repairing the damage caused by the use of sulphuric acid, resulting in fewer energy reserves for the embryo, consequently resulting in lower initial seedling growth in this treatment.

The majority of studies show the positive effects of chemical scarification with sulphuric acid on breaking the dormancy of Brachiaria sp. seeds. Despite this positive effect, care should be taken when using this method, as there is evidence that it causes damage to the seeds, which can interfere with the initial development of plants.

Potassium nitrate (KNO_3)

According to the Rules for Seed Analysis (RAS) (BRASIL, 2009), an aqueous solution containing 0.2% potassium nitrate (KNO_3) can be used to moisten the substrate used in the germination test. However, Costa et al. (2011) found that potassium nitrate had no positive effect on breaking the dormancy of Brachiaria humidicola seeds, whether or not they were scarified with sulphuric acid.

Binotti et al. (2014) evaluated different periods (0, 2, 4, 6 and 8 hours) of pre-soaking Brachiaria brizantha cv. MG-5 seeds in potassium nitrate solution (0.2%) and verified the germination percentage on different dates after the experiment was set up (7, 14 and 21 days).

The researchers observed significant differences, with a quadratic model (Y = 43.785714 + 3.6642857X - 0.65178571X2 and R2 = 0.92) only in the first evaluation period (7 days).

Binotti et al. (2014) also found that by increasing the soaking time of the seeds, there was a linear reduction (Y = 10.375006 - 0.0282803x and R2 = 0.97) in the number of viable (dormant) seeds remaining from the germination test, showing that soaking the seeds in KNO3 is an interesting pre-germination process for overcoming dormancy.

In addition to its positive effect on seed germination, Costa et al. (2011) report that KNO3 provides a higher percentage of normal seedlings than hydrogen peroxide (H2O2).

Also, for the germination speed index (GVI), Binotti et al. (2014) observed a higher value (3.95) with a pre-soaking time of up to 3.4 hours in a solution containing water + KNO3, showing that keeping the seeds for a certain period in a solution containing water + KNO3 made the Brachiaria seed coat more permeable, facilitating soaking and speeding up the germination process.

Batista et al. (2015), on the other hand, evaluated the use of potassium nitrate in the first hydration cycle, which had no influence on the germination patterns of Brachiaria brizantha cv. MG-5 seeds, compared to the use of gibberellin and water (control). However, the researchers observed that the use of potassium nitrate in the first hydration cycle, combined with a second cycle using water and calcium nitrate, resulted in the repair of the membrane system of Brachiaria brizantha seeds.

Although Batista et al. (2015) found no effect of KNO3 on breaking the dormancy of Brachiaria brizantha cv. MG-5, other studies point to a positive effect of this substance (COSTA et al., 2011), (BINOTTI et al., 2014), justifying its use in Brachiaria sp. seeds.

Boiling water

The use of water stands out as a method that is easy to apply and has the potential to be used to break dormancy in some species. According to Fowler and Bianchetti (2000), the use of cold water (soaking in water at 25 °C for 24 hours) is recommended for species that have difficulties germinating, even if they are not dormant. These problems stem from long periods of storage, which cause the seeds to dry out, initially making it difficult to absorb water and germinate.

Hot water immersion, on the other hand, is effective at overcoming seed tegument dormancy, but the temperature and time vary between species (FOWLER and BIANCHETTI, 2000). Silva et al. (2014) evaluated seeds from three cultivars of Brachiaria sp., which were immersed in water heated to 60 °C for 5 minutes, and found no difference in germination rate between the cultivars, nor any difference between the treatment using H_2SO_4, KNO3 and water (control).

For Brachiaria brizantha cv. Marandu seeds, Lacerda et al. (2010) reported that the use of boiling water is an excellent alternative for overcoming dormancy. The researchers found that treatments using boiling water, regardless of how long the seeds were immersed (1, 2 or 3 minutes), showed lower levels of dormancy and were more effective than the use of gibberellic acid (100 and 200 mg L^{-1} - for 48 hours), scarification with concentrated sulphuric acid and the control treatment.

Studies on forest species have demonstrated the positive effects of using hot water to break dormancy. However, there are few recent studies in the literature on the use of hot water as a method for breaking dormancy in Brachiaria sp.

Storage time

With regard to storage time, Costa et al. (2011) observed that Brachiaria humidicola (Rendle) Schweick seeds reduce dormancy over time, and that storing the seeds for 21 months resulted in fewer dormant seeds compared to sulphuric acid scarification and the application of germination-promoting substances (potassium nitrate (0.2%); gibberellic acid (100 mg L^{-1}) and; hydrogen peroxide (0.30%)).

Dias and Alves (2008) also evaluated different batches of Brachiaria brizantha cv. Marandú seeds and observed the influence of storage time on seed germination, with most of the batches evaluated reducing the number of dormant seeds over time.

Alves et al. (2017) found that storage time positively influences the germination of forage seeds. In addition to storage time, Cardoso et al. (2014) observed that the accelerated ageing technique also overcomes dormancy and stimulates greater germination of Brachiaria brizantha cv. MG-5 Vitória seeds.

These studies have shown that Brachiarias sp. have seed dormancy immediately after harvest, which is reduced over the course of storage.

Germination temperature

According to Fowler and Bianchetti (2000), seeds enter embryonic dormancy when subjected to unfavourable conditions for germination, such as high or low temperatures.

In order to identify whether temperature has an effect on the germination of Brachiaria seeds, Chiodini and Cruz-Silva (2013) evaluated the effect of different temperatures (10, 25 and 40 °C) on the germination of Brachiaria brizantha cv. Marandu (Hochst. ex A. Rich.) Stapf seeds, and concluded that a temperature of 25 °C is more suitable, showing greater germination.

FINAL CONSIDERATIONS

Based on research in the literature, it can be seen that there are various methods that can be used alone or in combination to reduce the percentage of dormant seeds in Brachiaria sp.

Practically all the methods showed positive results, but opposite effects were also observed, with problems occurring in the development of the plants in sequence.

In addition, there are some differences between some studies, which indicates the possibility of more than one dormancy mechanism acting on Brachiaria sp. seeds at the same time.

BIBLIOGRAPHICAL REFERENCES

ALVES, B.A.; MEDEIROS, L.T.; SALES, J.F.; BRANQUINHO, A.C.; SILVA, J.W.; SOUZA, R.G. Germination of forage seeds of the genus Brachiaria as a function of environments and storage times. Global Science and Technology, Rio Verde, v.10, n.01, p.11 - 19, 2017.

BATISTA, T.B.; CARDOSO, E.D.; BINOTTI, F.F.S.; SÁ, M.E.; HAGA, K.I. Nutrients and gibberellin in physiological conditioning on brachiaria seed quality. Revista de Agricultura Neotropical, Cassilândia- MS, v. 2, n. 1, p. 10-16, jan./mar. 2015.

BINOTTI, F.F.S.; SUEDA JUNIOR, C.I.; CARDOSO, E.D.; HAGA, K.I.; NOGUEIRA, D.C. Pre-germination treatments in Brachiaria seeds. Agrária - Revista Brasileira de Ciências Agrárias, v.9, n.4, p.614-618, 2014. DOI:10.5039/agraria.v9i4a2781

BRAZIL. Ministry of Agriculture, Livestock and Supply. Secretariat for Agricultural Defence. Rules for analysing seeds. 2009.

CARDOSO, E.D.; SÁ, M.E.; HAGA, K.I.; BINOTTI, F.F.S.; NOGUEIRA, D.C.; VALÉRIO

FILHO, W.V. Physiological performance and overcoming dormancy in Brachiaria brizantha seeds submitted to chemical treatment and artificial ageing. Semina: Agricultural Sciences, v. 35, n. 1, 2014. DOI: 10.5433/1679-0359.2014v35n1p21

CHIODINI, B.M.; CRUZ-SILVA, C.T.A. Effect of temperature on seed germination of Brachiaria brizantha cv. Marandu (Hochst. ex A. Rich.) Stapf (Poaceae). Varia Scientia Agrárias, v. 3, n. 2, p. 105-113, 2013.

COSTA, C.J.; ARAÚJO, R.B.; BÔAS, H.D.C.V. Treatments for overcoming dormancy in Brachiaria humidicola (Rendle) Schweick seeds. Pesquisa Agropecuária Tropical, v. 41, n. 4, 2011.

DIAS, M.C.L.L.; ALVES, S.J. Evaluation of the viability of Brachiaria brizantha (Hochst. ex A. Rich) Stapf seeds by the tetrazolium test. Revista Brasileira de Sementes, v. 30, n. 3, p. 145-151, 2008. DOI: 10.1590/S0101- 31222008000300019

FABRICE, C.E.S.; SOARES FILHO, C.V.; PINTO, M.F.; PERRI, S.H.V.; CECATO, U.; MATEUS, G.P. Recovery of degraded "Brachiaria decumbens" pastures with the introduction of Stylosanthes and phosphate fertilisation. Brazilian Journal of Animal Health and Production, v. 16, n. 4, p. 758-771, 2015.

FOWLER, J.A.P.; BIANCHETTI, A. Dormancy in forest seeds. Colombo: Embrapa Florestas, 2000.

GARCIA, C.M.D.P.; ANDREOTTI, M.; TEIXEIRA FILHO, M.C.M.; BUZETTI, S., CELESTRINO, T.S.; LOPES, K.S.M. Agronomic performance of the maize crop and forage species in a Crop-Livestock Integration system in the Cerrado. Ciência Rural, v. 43, n. 4, 2013.

LACERDA, M.J.R.; CABRAL, J.S.R.; SALES, J.F.; FREITAS, K.R.; FONTES, A.J. Overcoming seed dormancy in *Brachiaria brizantha* cv. "Marandu". Semina: Ciências Agrárias, v. 31, n. 4, 2010.

PETTER, F.A. PACHECO, L.P.; ZUFFO, A.M.; PIAUILINO, A.C.; XAVIER, Z.F.; SANTOS, J.M.; MIRANDA, J.M.S. Performance of cover crops submitted to water deficit. Semina: Agricultural Sciences, v. 1, n. 34, 2013. DOI 10.5433/1679-0359.2013v34n6Supl1p3307

SCHEIDT, K.C.; MESQUITA, E.E.; OLIVEIRA, N.T.E.; MECABÔ, C.A.; LIPSCH, R.R.; CASTAGNARA, D.D.; CRISTOFORI, E.C.; GIARETTA, L.E. Morphogenic characteristics of *Brachiaria ruziziensis* in intercropping systems with maize. Scientia Agraria Paranaensis, v. 13, p. 339-342, 2014. DOI: http://dx.doi.org/10.1818/sap.v13i0.11172

SILVA, A.B.; LANDGRAF, P.R.C.; MACHADO, G.W.O. Germination of brachiaria seeds under different concentrations of gibberellin. Semina: Agricultural Sciences, v. 34, n. 2, 2013. DOI: 10.5433/1679-0359.2013v34n2p657

SILVA, A.L.; TORRES, F.E.; GARCIA, L.L.; MATTOS, E.M.; TEODORO, P.E. Treatments for breaking dormancy in *Brachiaria brizantha*. Revista de Ciências Agrárias, v. 37, n. 1, p. 37-41, 2014.

ZUFFO, A.M.; ANDRADE, F.R.; SILVA, L.M.A.; MENEZES, K.O.; SILVA, R.L.; PIAUILINO, A.C. Depth of sowing and overcoming dormancy in the initial growth of *Brachiaria dictyoneura* (Fig. & De Not.). Stapf (1919) cv. Llanero. Revista Ceres, Viçosa, v. 61, n.6, p. 948-955, 2014. DOI: http://dx.doi.org/10.1590/0034-737X20146106009

CHAPTER 4

PRODUCTIVITY OF CROP CORN IN MONOCULTURE AND CONSORTIUM WITH Urochloa brizantha AND Cajanus cajan

Lucas Gonçalves de Oliveira
Paulo Roberto Rabelo
Lucas Link
Ana Claudia Schllemer dos Santos
Karine Fuschter Oligini
Vanderson Vieira Batista
Paulo Fernando Adami

Summary: The intercropping of maize with other species may have the potential to increase the amount of residual straw on ploughed areas. However, research has also shown that cereal yields can be compromised in this system. In this context, the aim of the study was to assess the yield components of the maize crop, grown in consortium with Urochloa brizantha (brachiaria) and Cajanus cajan (guandu beans), and compare them with the single crop system. The study was conducted in the municipality of Dois Vizinhos, Paraná, during the 2016/2017 harvest. The treatments evaluated were maize grown in monoculture, maize intercropped with Brachiaria brizantha and maize intercropped with Cajanus canjan. Plant height, ear insertion height, maize population, number of grains per row and row per ear, number of grains per ear, thousand grain mass and maize yield were assessed. The data was subjected to analysis of variance (ANOVA) ($p > 0.05$) and if significant, Tukey's test of means was applied at 5% probability. Maize grown in a consortium with Urochloa Brizantha had lower ear insertion height, number of grains per row, thousand grain mass and yield compared to maize grown alone. Maize intercropped with Cajanus cajan resulted in a lower final plant population (67,406 plants ha^{-1}) and lower productivity (13,433 Kg ha^{-1}) than maize grown alone (13,868 Kg ha^{-1}). Intercropping maize with Cajanus cajan also resulted in higher yields than intercropping with Urochloa Brizantha (13,011 Kg ha^{-1}). The intercropping system still needs to be studied in order to identify the appropriate management for the system so that there is no reduction in maize yields.

INTRODUCTION

Current maize hybrids are more tolerant of plant density (TOLLENAAR, 1992). Although maize is highly productive, Andrade (1995) points out that the cultivation and management of the cereal must be planned in such a way as to minimise climatic factors, which cannot be controlled and can have an impact on its productive capacity.

Growing maize during the off-season is becoming an option for producers in the state of Paraná, due to the soya crop's sanitary void, which does not allow it to be planted during this period, and the area cultivated with maize during this period is increasing (CONAB, 2016). Intercropping in Brazil has been gaining significant ground, with data showing that pastureland occupies 48 per cent of the country's territory (IBGE, 2006). As a result, the use of integrated cultivation of annual crops and pastures is a tool for recovering degraded soils, as well as benefiting subsequent crops (SEVERINO et al., 2006).

The establishment of maize in consortium with different cover crops is an alternative that aims to maximise the use of the area, making the most of the nutrients present in the soil. Integrated cultivation has been suggested as an alternative for recovering degraded soils and benefiting subsequent crops (SEVERINO et al., 2006), making it an economically viable technique (GARCIA et al., 2012). This practice also helps to reduce weed emergence (ARAÚJO et al., 2000), (SEVERINO et al., 2006), (JAKELAITIS et al., 2003), (MATEUS et al., 2004), (TREZZI and VIDAL, 2004).

Several studies have already been carried out on the intercropping method with the aim of understanding the system even more. Pequeno et al. (2006) evaluated sowing time and herbicide management. Freitas et al. (2008), Ferreira et al. (2005) and Jakelaitis et al. (2005) studied the use of different forage species and Ceccon (2008) highlighted the use of tropical forage species such as Urochloa braizantha alongside maize. Also, according to Giacomini et al. (2003), a consortium of cover crops not only helps to protect the soil, but can also provide nitrogen via biological fixation.

Despite all the benefits of intercropping maize with cover crops, maize yields can be negatively affected by this system. Borghi and Crusciol (2007) concluded that when 45 cm spacing is used, maize grain yield is lower when B. brizantha is intercropped in the row and between the rows.

In this context, the aim of the study is to assess the yield components of the maize crop, grown in different production systems, consortium between maize + Urochloa brizantha (brachiaria), maize + Cajanus cajan (guandu beans), comparing it with the single crop system.

MATERIAL AND METHODS

The study was conducted at Cooperativa Agroindústria COASUL, in Dois Vizinhos, Paraná, latitude 25°43'54.33" South, longitude 53°05'5.66" West and an altitude of 606 metres. The region's climate is classified as Cfa (ALVARES et al. 2013), with rainfall between 1,800 and 2,000 mm (IAPAR, 2016) and dystrophic red latosol (BHERING et al., 2008). The experimental area is cultivated in a no-till system and has a history of experiments conducted by the co-operative, being used for field days and evaluating the potential of materials.

A randomised block design was used, with three treatments and four replications. The treatments consisted of different cropping systems: maize grown in monoculture, maize intercropped with Brachiaria brizantha and maize intercropped with Cajanus canjan.

The experimental units (EU) were made up of 10 rows of maize, spaced 45 cm apart and 20 metres long. Evaluations were carried out on the sampling units (AU), which consisted of the three central rows, disregarding the first and last linear metre of these rows, generating AUs of 24.3 m^2 .

The study used the P30F53 YHR hybrid, which was sown on 9 September 2016 at a density of 70,000 seeds ha^{-1} . The cover crops were sown simultaneously, between the corn rows, at a density of 25 kg ha^{-1} for guandu beans and 13 kg ha^{-1} for Urochloa brizantha. Fertilisation in the sowing furrow was carried out with 500 Kg ha^{-1} of the 08-20-20 formulation. Two top dressings were also made with 330 Kg ha^{-1} of urea (45% nitrogen) each, the first at V3 and the second at V8.

The experimental area was desiccated with the herbicide Roundup Original® (Glyphosate) 900 g ha^{-1} , with 200 L ha^{-1} of mixture to control the weeds. At the V3 phenological stage, atrazine (3.5 kg i.a ha^{-1}) was applied in an attempt to achieve good weed management in the single maize and maize + Urochloa brizantha areas. In the maize + guandu area, weed control was carried out manually as there were no selective herbicides for guandu beans. There was also an application of the fungicide Abacus 0.4 L ha-1, applied at V8, with a backpack machine and a spray volume of 200 L ha^{-1}.

Evaluations were carried out at the VT phenological stage (vegetative stage) of the maize crop, and the following variables were assessed: plant height (m), ear insertion height (m), population (plants ha^{-1}). Plant height and ear insertion height were obtained from 10 plants per AU. Plant height was taken to mean the distance between the base of the soil and the tip of the shoot, and ear insertion height was taken to mean the distance between the soil and the base of the ear. The population was determined by counting the number of plants present in each AU, and the values were extrapolated to hectares.

When the maize reached grain moisture of approximately 22%, the harvest and yield component evaluations were carried out. Ten random ears per AU were assessed for the number of rows per ear and the number of grains per row, and the average value observed was used to analyse the data. Also, by multiplying the values for rows per ear and grains per row, the number of grains per ear was determined.

The mass of one thousand grains was obtained from three samples of one hundred grains, which were weighed and the value corrected to 13% moisture and multiplied by 10. Yield was determined by threshing all the ears present in each AU, the grain sample being weighed, the moisture corrected to 13% and extrapolated to the hectare.

The data was subjected to analysis of variance (ANOVA) ($p > 0.05$) and if significant, Tukey's test of means at 5% probability was applied using the Statgraphic Plus 4.1 programme (MANUGISTICS, 1997).

RESULTS AND DISCUSSIONS

The 2016/2017 harvest was characterised as an atypical year for agriculture in Paraná, as the climate was favourable with regular (Figure 1 A) and well-distributed rainfall followed by satisfactory temperatures (Figure 1 B) and luminosity, meeting the needs of the maize crop during the study period.

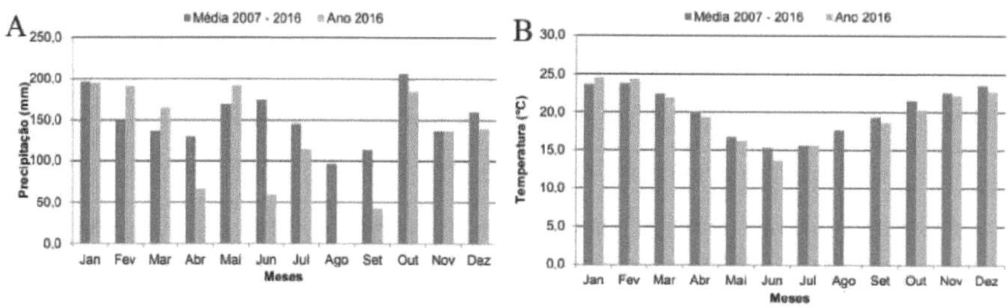

Figure 1. Precipitation (mm) (A) and average temperature (°C) (B) from 2007 to 2016 and in 2016 (GEBIOMET, 2016). Source: Author, 2018.

There was no difference between the treatments for the variable plant height, with an average of 2.61 cm (Table 1). However, for ear insertion height, the single maize cropping system with values of 1.66 m stood out, differing from the cropping system consisting of maize intercropped with Urochloa brizantha (1.52 m), but being similar to the intercrop with Cajanus cajan (1.61 m) (Table 1).

Table 1. Plant height, ear insertion height and plant population of maize grown in consortium with Urochloa brizantha (brachiaria), Cajanus cajan (guandu beans) and maize in monoculture (maize). Dois Vizinhos, Paraná, 2018.

Treatments	Plant height (m)	Ear insertion height (m)	Population (plants ha)$^{-1}$
Maize + brachiaria	2,55^{ns}	1,52 b	68,147 ab
Maize + guandu beans	2,63	1.61 ab	67.406 b
Maize	2,64	1,66 a	69.628 a
Average	2,61	1,60	68.394

^{ns} not significant. Averages with different letters in the column differ by Tukey's test at 5% probability.

The fact that there was no statistical difference between the treatments in terms of plant height suggests that the species used in the consortium did not interfere with the development of the maize plants. According to Freitas et al. (2008), corn intercropping is viable because the crop develops quickly and its development is not affected. Although U. brizantha reduces the ear insertion height, the value of 1.52 metres is considered satisfactory, as it still allows mechanical harvesting and is not an obstacle to implementing the system. Ear height of less than 1.2 m can affect mechanised harvesting, causing harvest losses and increasing impurities in the grain (POSSAMAI et al., 2001).

In terms of plant population, the single maize treatment, with 69,628 plants ha^{-1} , differed from the consortium with guandu beans, which had 67,406 plants ha^{-1} (Table 1). However, there was no difference in plant population between the maize + U. brizantha intercropping system and the others (Table 1). The lower maize population in the consortium is possibly related to competition between the plants for water, light and nutrients.

The number of rows per ear is an important parameter in the production potential of the maize crop. In this study, this variable did not differ statistically between treatments (Table 2), indicating that the treatments did not influence this factor. According to Ritchie et al. (1993) the number of rows is a parameter defined between phenological stages V4 and V8. In other words, during this stage of maize development, there was no competition to the point of affecting this yield component.

Table 2. Rows per ear, grains per row, grains per ear, thousand grain mass and yield of maize grown in consortium with Urochloa brizantha (brachiaria), Cajanus cajan (guandu beans) and maize in monoculture (maize). Dois Vizinhos, Paraná, 2018.

Treatments	Rows per cob	Grain per row	Grain per cob	Mass of thousand grains (g)	Productivity (Kg ha)$^{-1}$
Maize + brachiaria	15,60^{ns}	38,46 b	600^{ns}	395 b	13.011 c
Maize + guandu beans	15,73	41,13 a	647	425 a	13.433 b
Maize	15,46	42,40 a	655	435 a	13.868 a
Average	15,60	40,66	634	418	13.437

^{ns} not significant. Averages with different letters in the column differ by Tukey's test at 5% probability.

Table 2 shows that the number of grains per row was lower when corn and brachiaria were intercropped, but this reduction did not influence the number of grains per ear, which was similar between the treatments studied (Table 2). The mass of a thousand grains was also lower when maize was intercropped with Urochloa brizantha (Table 2).

As for productivity, Table 2 shows that when maize was grown alone, its productivity was higher than when it was intercropped. With a yield of 13,433 Kg ha^{-1} , it was found that the intercropping of maize with guandu beans was higher than the yield of maize intercropped with brachiaria, which was only 13,011 Kg ha^{-1} (Table 2). These results differ from those of Kluthcouski et al. (2000) and Freitas et al. (2005), who reported that maize productivity in intercropping with U. brizantha is not affected compared to maize cultivation alone.

According to Borgui et al. (2006), the fact that maize and brachiaria are grasses can increase the demand for nitrogen in the system, which is greater than when legumes are used. Therefore, the U. brizantha maize consortium tends to produce less than the maize and legume consortium.

Although intercropping has reduced maize yields, it has the potential to be used on rural properties, as there is a greater accumulation of biomass in the intercropping system, as the straw contributes to soil protection (GIACOMINI et al., 2003), weed reduction (ARAÚJO et al., 2000), (JAKELAITIS et al., 2003) and (MATEUS et al., 2004), in the recovery of degraded soils (SEVERINO et al., 2006), and can be economically viable (GARCIA et al., 2012). The challenge still facing intercropping is to manage the crops properly in order to reduce competition between the species, so that the maize crop does not reduce grain yields.

CONCLUSIONS

Maize grown in a consortium with Urochloa Brizantha reduces ear insertion height, the number of grains per row, thousand grain mass and yield compared to maize grown alone.

Maize grown in consortium with Cajanus cajan reduces the final plant population and yield compared to maize grown alone.

Maize intercropped with Cajanus cajan results in higher yields than intercropped with Urochloa Brizantha.

BIBLIOGRAPHICAL REFERENCES

ALVARES, C. A.; STAPE, J. L.; SENTELHAS, P. C.; MORAES, J. L. G; SPAROVEK, G. Kõppen's climate classification map for Brazil. Meteorologische Zeitschrift, Stuttgart, v. 22, p.711-728, 2013.

ANDRADE, F. H. Analysis of growth and yield of maize, sunflower and soybean grown at Balcarce. Argentina. Field Crops Research, 41:1-12, 1995.

ARAÚJO, A. G.; RODRIGUES, B. N. Mechanical and chemical management of black oats and its influence on decomposition rate and weed control in direct maize sowing. Planta Daninha, v. 18, n. 1, p. 151-160, 2000.

BHERING, S. B., DOS SANTOS, H. G., BOGNOLA, I. A., CÚRCIO, G. R., CARVALHO JUNIOR, W. D., CHAGAS, C. D. S., SILVA, J. D. S. Soil map of the State of Paraná, updated legend. In: Embrapa Florestas-Article in conference proceedings (ALICE). In: BRAZILIAN CONGRESS OF SOIL SCIENCE, 32nd, 2009, Fortaleza. Soil and bioenergy production: perspectives and challenges: proceedings. [Viçosa, MG]: SBCS; Fortaleza: UFC, 2009., 2008.

BORGHI, É.; CRUSCIOL, C. A. C.; COSTA, C. Development of the maize crop in intercropping with Brachiaria brizantha in a no-till system. Engenharia Agrícola, Botucatu, v. 21, n. 3, p.19-33, dec. 2006.

BORGHI, E.; CRUSCIOL, C A. C. Maize productivity, spacing and intercropping with Brachiaria brizantha in a no-till system. Pesquisa Agropecuária Brasileira, p. 163-171, 2007.

CECCON, G. Maize with brachiaria in consortium. Technical communiqué, 140. Dourados: Embrapa Agropecuária Oeste, 2008.

CONAB. NATIONAL SUPPLY COMPANY. Monitoring the Brazilian grain harvest. Available at:< http://www.conab.gov.br/OlalaCMS/uploads/arquivos/16_11_11_14_54_21_boletim_graos_novembro_2016.pdf>. Accessed on: 16 November 2016.

FERREIRA, A. C. S. L. R., SILVA, A. A., SEDIYAMA, A. F. B. C. S. Morphological characteristics of soya and brachiaria intercropped under subdoses of fluazifop-p-butil. Ciência Rural, Santa Maria, v. 35, n. 2, p.277-283, Apr. 2005.

FREITAS, F. C. L., FERREIRA, L. R., FERREIRA, F. A., SANTOS, M. V., AGNES, E. L., CARDOSO, A. A., JAKELAITIS, A. Pasture formation via consortium of Brachiaria brizantha with maize for silage in the no-tillage system. Planta Daninha, v. 23, n. 1, p. 49-58, 2005.

FREITAS, F. C. L.; SANTOS, M. V., MACHADO, A. F. L., FERREIRA, L. R., FREITAS, M. A. M., SILVA, M. G. O. Behaviour of maize cultivars in intercropping with Brachiaria brizantha in the presence and absence of foramsulfuron + iodosulfuronmethyl for forage management. Planta Daninha, Viçosa, v. 26, n. 1, p. 215-221, jan./mar. 2008.

GARCIA, C. M. de P.; ANDREOTTI, M.; TARSITANO, M. A. A.; TEIXEIRA FILHO, M. C. M.; LIMA, A. E. da S.; BUZETTI, S. Economic analysis of the grain yield of maize intercropped with forage plants of the genera Brachiaria and Panicum in a no-till system. Revista Ceres, Viçosa, MG, v. 59, n. 2, p. 157-163, 2012.

GEBIOMET, Biometerology study group. Available at: < http://www.gebiomet.com.br/>. Accessed on: 15 January 2018.

GIACOMINI, S. J., AITA, C., VENDRUSCOLO, E. R. O., CUBILLA, M., NICOLOSO, R. S., FRIES, M. R. Dry matter, C/N ratio and accumulation of nitrogen, phosphorus and potassium in mixtures of ground cover plants. Revista Brasileira de Ciência do Solo, Brasília, v. 27, p. 325-334, 2003.

IAPAR, Agronomic Institute of Paraná. Paraná agroclimatic monitoring system. 2016. Available at : <http://www.iapar.br/modules/conteudo/conteudo.php?conteudo=595>. Accessed 26 Oct. 2017.

IBGE. Agricultural census 1920/2006. Until 1996, data extracted from: Statistics of the 20th Century. Rio de Janeiro: IBGE, 2006. Available at:<http://seriesestatisticas.ibge.gov.br/>. Accessed on: 10 Nov. 2016.

JAKELAITIS, A. FERREIRA, L. R., SILVA, A. A. D., AGNES, E. L., MIRANDA, G. V., MACHADO, A. F. L. Population dynamics of weeds under different management systems in maize and bean crops. Planta Daninha, v. 21, n. 1, p. 71-79, 2003.

JAKELAITIS, A., SILVA, A. A., FERREIRA, L. R., SILVA, A. F., PEREIRA, J. L., VIANA, R. G. Effects of herbicides on maize intercropped with Brachiaria brizantha. Planta Daninha, Viçosa, v. 23, n. 1, p. 69-78, jan./mar. 2005.

KLUTHCOUSKI, J., COBUCCI, T., AIDAR, H., YOKOYAMA, L. P., DE OLIVEIRA, I. P., COSTA, J. D. S., ... MAGNABOSCO, C. D. U. Sistema Santa Fé-Tecnologia Embrapa: integração lavoura-pecuária pelo consórcio de culturas anuais com forrageiras, em áreas de lavoura, nos sistemas direto e convencional. Embrapa Rice and Beans. Technical circular, p.28, 2000.

MANUGISTICS. Statgraphics plus for Windows (version 4.1). Rockville, Maryland, 1997. CD-ROM.

MATEUS, G. P.; CRUSCIOL, C. A. C.; NEGRISOLI, E. Giant guinea sorghum straw on weed establishment in a no-till area. Pesquisa Agropecuária Brasileira, v. 39, n. 6, p.

539-542, 2004.

PEQUENO, D. N. L., MARTINS, E. P., AFFERRI, F. S., FIDELIS, R. R., & SIQUEIRA, F. L. T. D. Effect of sowing time of Brachiaria brizantha in consortium with maize, on agronomic characters of the annual crop and the forage in Guripi, State of Tocantins. Amazônia: Ciência e Desenvolvimento, Belém, v. 2, n. 3, p. 127-133, Jul./Dec., 2006.

POSSAMAI, J.M. et al. Soil preparation systems for safrinha maize cultivation. Bragantia, v.60, n.2, p.79-82, 2001.

RITCHIE, S. W.; HANWAY, J. J.; BENSON, G. O. How a corn plant develops. Ames: Iowa State University of Science and Technology, 1993. 26p. (Special report, 48).

SEVERINO, F. J.; CARVALHO, S. J. P.; CHRISTOFFOLETI, P. J. Mutual interference between maize, forage species and weeds in a consortium system. II - implications for forage species. Planta Daninha, v. 24, n. 1, p. 45-52, 2006.

SEVERINO, F. J.; CARVALHO, S. J. P.; CHRISTOFFOLETI, P. J. Mutual interference between maize, forage species and weeds in a consortium system. III - implications for weeds. Planta Daninha, v. 24, n. 1, p. 53-60, 2006.

TREZZI, M. M.; VIDAL, R. A. Potential use of sorghum and millet mulches in weed suppression under field conditions: II - Effects of mulching. Planta Daninha, v. 22, n. 1, p. 1-10, 2004.

TOLLENAAR, M. Is low plant density a stress in maize? Maydica, v.37, p.305-311, 1992.

CHAPTER 5

PERFORMANCE OF SOYA CULTIVARS AND HARVEST LOSSES

Lucas Link
Ana Claudia Schllemer dos Santos
Eduardo Júnior de Oliveira da Silva
Paulo Roberto Rabelo
Vanderson Vieira Batista
Karine Fuschter Oligini
Paulo Fernando Adami

Summary: The global demand for soya beans is increasing, which requires new techniques to reduce harvest losses, as well as cultivars that adapt to different growing seasons, in order to increase national soya bean production. The aim of this study was to assess the development and productivity of four soya cultivars and to check their relationship with harvest losses. The study was conducted in the municipality of Dois Vizinhos - PR, in a randomised block design with four replications. The soya cultivars used were: Agroeste 3610 Intacta RR2, Agro Norte 78017 (Mato Grosso), Brasmax Força RR and Coodetec 205, both commercial. At the end of each cultivar's cycle, the following were assessed: number of internodes, plant height, height of first pod insertion, number of grains, number of pods, thousand-grain mass, yield and harvest losses. A Tukey mean comparison test and correlation test between variables were carried out. The most productive cultivar was Brasmax Força RR (2,885.5 kg ha^{-1}), followed by the cultivars Coodetec 205 (2,325.9 Kg ha^{-1}), Agroeste 3610 Intacta RR2 (1,778.3 Kg ha^{-1}) and Agro Norte 78017 (1,088.1 Kg ha^{-1}). A loss higher than the acceptable level was observed for the Coodetec 205 cultivar, 88 Kg ha^{-1} , but the losses did not correlate with any of the cultivars' morphological components.

INTRODUCTION

The soya bean (Glycine max) is the most widely consumed and produced legume in the world, and among grains it is the commodity that generates the most commercial income worldwide. In addition, its grains are used both in human nutrition, whether fresh or in the form of oil, and in animal nutrition, both fresh and in feed formulation (CONTINI et al., 2014), as well as having energy potential for biodiesel production.

Due to its widespread use, soya began to be grown in Brazil in the 1930s, in the south of the country, as an option for summer cultivation, since wheat was the main crop in winter. The cultivation of soya began slowly, with varieties brought from the United States of America. However, over the years, and together with the need for raw materials to feed animals such as pigs and poultry, soya began to be cultivated intensively, using technologies to increase its productivity (BONATO et al., 2002).

Brazil is currently the world's second largest soya producer, with production in the 2016/17 harvest of 97 million tonnes, behind the United States of America (USA), which produces 107.1 million tonnes. In Brazil, Mato Grosso is the largest producer, with Paraná in second place in the soya production ranking (SEAB, 2017).

In order to become the world's largest soya producer, it is necessary to invest in genetics, fertilisation, mechanisation and new ways of growing soya (NEVES, 2011). However, one problem observed in soya cultivation is harvesting losses, especially in the cutting bar, due to very low pod insertion, which combined with the volume of straw from previous crops, such as maize, can hinder the harvesting process, generating significant losses (SCHANOSKI et al., 2011).

According to Smiderle et al. (2009), one way to reduce harvest losses is to be careful when sowing, avoiding sowing in conditions of high humidity and at high speed, as these factors increase the opening of the furrow, forming a deeper furrow, and consequently the first pod of the plant is very close to the ground. Likewise, the cultivars to be used should be analysed, opting for cultivars that are productive and at the same time have a higher first pod insertion, favourable plant height and number of branches, thus reducing harvest losses.

The aim of this study was therefore to assess the development and productivity of four soya cultivars and check their relationship with harvest losses.

MATERIALS AND METHODS

The experiment was carried out in an agricultural area on the Link Estate, located in the community of Santa Maria, Dois Vizinhos, PR. The soil is classified as Latossolo Vermelho Distroférrico, and its chemical and physical analyses are shown in Table 1.

Table 1. Chemical and physical analysis of the experimental area. Dois Vizinhos, PR, 2018.

K	Ca	Mg	Al	H+Al	CTC	SB	pH
			$cmol_c . dm^{-3}$				CaCl2
0,46	9,35	2,88	0,0	6,21	18,9	12,69	5,0
					Particle size analysis		
MO		P		V	Clay	Sand	Silt
g.dm-3		$mg.dm^{-3}$			%		
26,42		2,17		67,14	51,25	20,0	28,75

The area where the experiment was carried out has been no-till for 18 years, with crop rotation in winter (wheat, oats, triticale, turnip rape and ryegrass) and summer (maize, soya, sorghum and beans).

The soya cultivars were sown in a randomised block design with four replications. Sowing was carried out on 19 and 20 February 2014, in the recently harvested maize crop remains. Sowing was carried out using a New Holland tractor, model TM 7010 (NEW HOLLAND, 2017a) and a John Deere seeder, model 1109 (JOHN DEERE, 2014), with a VacuMeter seed distribution system™ .

The seeder used has nine pantograph rows spaced 45 cm apart. For a smaller furrow opening, the seed drill used a smooth cutting disc, a double offset disc for fertiliser and a double offset disc for seed. The seed metering system was of the VacuMeter type™ , with a vertically arranged disc, 108 holes and operating at a negative pressure of 12 bar. The sowing speed was 6.5 km h^{-1} .

Each plot consisted of 18 crop rows 50 metres long. Two evaluation units were previously established in each plot, one to evaluate the soya components and the other to determine harvest losses, which had dimensions of 5.61 x 0.5 m, i.e. 2.8 m^2 .

The soya cultivars used in the study were Brasmax Força RR (BRASMAX, 2015), Agroeste 3610 Intacta RR2 PRO (AGROESTE, 2015), Coodetec 205 (COODETEC, 2010) and Agro Norte 78017. Sowing consisted of depositing 21 seeds per linear metre to give a population of 467,000 plants per hectare. The base fertiliser was 165 kg ha^{-1} of NPK 02-20-10.

Pests, diseases and weeds that appeared in the crop were managed according to the needs of the cultivars. Pre-sowing desiccation used the herbicides Glyphosate + Chlorimuron. For post-sowing desiccation, only the herbicide Glyphosate was used. Three applications of fungicide and insecticide were made, the first with the fungicide Picoxystrobin + Cyproconazole and the insecticides Imidacloprid + Beta-Cyfluthrin and Triflumuron. The second was the fungicides Picoxystrobin + Cyproconazole and Copper Oxychloride and the insecticides Thiamethoxam + Lambda Cyhalothrin and Teflubenruzon. The third was the fungicides Trifloxystrobin + Cyproconazole and Copper Oxychloride and the insecticide Acephate.

To analyse the performance of the cultivars, the number of internodes, pods per plant, grains per plant, height of insertion of the first pod and plant height (yield components) were assessed on ten plants from each sampling unit at the R8 stage. A millimetre tape measure was used to measure plant height and first pod insertion height. The number of internodes, pods per plant and grains per plant were counted manually.

The plants from each sampling unit were collected and their grains threshed using a grain whisk. The grain sample obtained was weighed, its moisture measured and corrected to 14% (RIBEIRO and BIAGGIONI, 2006), and the value extrapolated to hectares (kg ha^{-1}).

The mass of grains (mass of 1000 grains) (g) was determined by weighing 100 soya beans 10 times in each sample, adding up the values obtained and correcting the moisture content to 14% (RIBEIRO and BIAGGIONI, 2006).

The experimental plots were then harvested using a New Holland harvester, model TC 5070 Hydro (NEW HOLLAND, 2017b), with a 17-foot (5.61 m) cutting platform, a self-levelling sieve system (corrects the slope by up to 23%) and working at a speed of 4.5 km h^{-1}.

In order for the harvester to perform well, the harvesting settings were made, measured and noted down. The characteristics are as follows: cutting height 5 cm, cylinder speed 800 rpm, concave opening 2 cm at the front and 0.8 cm at the back, beater speed 875 rpm, rotary speed 400 rpm and chaff bag speed 212 rpm; sieve opening 3 in the middle and 4 on the sides, fan speed 750 rpm and windlass speed 23 rpm.

The grains lost by the harvester in each sampling unit were collected manually and stored in a plastic bag. Afterwards, the samples were taken to the Seed Analysis Laboratory at UTFPR Campus Dois Vizinhos, where they were weighed and the value transformed to hectare, with a humidity of 14% (RIBEIRO and BIAGGIONI, 2006).

The data was tabulated, an analysis of variance was carried out and a complementary analysis using Tukey at 1 and 5% probability, using Assistat statistical software (SILVA e AZEVEDO, 2002).

RESULTS AND DISCUSSION

According to the analysis of variance using the F test, the variables number of internodes, pods per plant, grains per plant and mass of 1000 grains were not different for the cultivars studied (Table 2). On the other hand, the height of insertion of the first pod and harvest losses differed at 5% probability, while plant height and yield differed statistically at 1% probability (Table 2).

Table 2. Analysis of variance using the F test for the variables number of internodes per plant (NE), number of pods per plant (VP), number of grains per plant (GP), height of insertion of the first pod (IV) in cm, plant height (AP) in cm. As well as the thousand grain mass (M1000) in grams, grain yield per area (PROD) in Kg ha^{-1} and grain losses per area (PER) in Kg ha^{-1} , Dois Vizinhos, PR, 2018.

	Elements assessed							
	Morphological characteristics					M1000	PROD	PER
Factors	NE	VP	GP	IV	AP			
F calc.	3,33^{ns}	1,68^{ns}	3,01^{ns}	9,21*	36,4**	2,32^{ns}	25,73**	4,07*
QM	2,18	328,34	2471,67	28,96	426,13	408,25	2.351.116,9	2.352,27

*, NE: number of internodes, VP: pods per plant, GP: grains per plant, IV: insertion of the 1ª pod, AP: plant height, M1000: mass of a thousand grains, PROD: grain yield, PER: grain losses.

Among the characteristics of the cultivars studied, no statistically significant differences were observed for the variable number of pods per plant (VP), with an average value of 40.5 pods per plant, for the variable number of internodes per plant (NE), which showed average values of 10.1 and for the number of grains per plant (95.6) (Table 3).

In terms of yield, the cultivar with the highest production was Brasmax Força RR, with 2,885.5 Kg ha^{-1} , followed by Coodetec 205, with 2,315.9 Kg ha^{-1} (Table 3), with a difference of 570 Kg ha^{-1} .

This implies that the Agro Norte 78017 cultivar is not adapted to the climatic conditions in the study period in the municipality of Dois Vizinhos, as it only produced 1,088 Kg ha^{-1} (Table 3), which is much lower than the production rates in Mato Grosso, where productivity reaches 3,396 Kg ha^{-1} (GARCIA et al., 2007). As soya is responsive to photoperiod, it may be that the Agro Norte 78017 cultivar did not adapt to the region, lengthening the cultivar's cycle and consequently making it more susceptible to the incidence of pests and diseases that attack soya at the end of the cycle, which can drastically reduce final productivity (GARCIA et al., 2007).

Table 3 Average morphological characteristics of soya cultivars: number of internodes per plant (NE), number of pods per plant (VP), number of grains per plant (GP), height of first pod insertion (IV) in cm, plant height (AP) in cm. As well as the thousand grain mass (M1000) in grams, grain yield per area (PROD) in kg ha^{-1} and grain losses per area (PER) in kg ha^{-1} , Dois Vizinhos, PR, 2018.

	Morphological characteristics					M1000	PROD	PER
Cultivars	NE	VP	GP	IV	AP			
Coodetec 205	9,9	49,2	123,8	6,1 b	49,1 c	143,6	2,315.9 ab	88,0 a
Brasmax RR Force	11,2	42,5	98,9	8,2 b	67,7 a	126,8	2.885,5 a	34,4 b
Agroeste 3610 Intact	9,6	27,7	63,2	12,3 a	58,9 b	122,5	1.778,3 b	47.3 ab
Agro Norte 78017	9,7	42,5	96,4	7,2 b	44,5 c	122,2	1.088,1 c	40.2 ab
Averages	10,1	40,5	95,6	8,4	55,1	128,8	2.016,9	52,5
CV%	8,0	34,6	29,9	20,9	6,22	10,3	15,0	45,8

*, NE: number of internodes, VP: pods per plant, GP: grains per plant, IV: insertion of the 1ª pod, AP: plant height, M1000: mass of a thousand grains, PROD: grain yield, PER: grain losses.

The cultivar Agro Norte 78017 had a yield 60.2 per cent lower than the Brazilian average, which is 2,732 Kg ha^{-1} (SEAB, 2017), and the cultivars Coodetec 205 and Agroeste 3610 Intacta had yields 15.2 per cent and 39.9 per cent lower than the Brazilian average, respectively. The only cultivar with yields above the national average was Brasmax Força RR (5.6%). The fact that most of the varieties have lower yields than the Brazilian average is due to the fact that they are planted in the late season, when there are lower temperatures, variations in humidity and, above all, shorter photoperiods (MARCHIORI et al., 1999) (CARVALHO et al., 2002).

According to Silveira and Conte (2013), plants with a height of less than 50 cm tend to have pod insertion closer to the ground, a fact that was verified in this experiment, when the cultivars Coodetec 205 and Agro Norte 78017 had plant heights of 49.1 and 44.5 cm, and first pod insertion of 6.1 and 7.2 cm, respectively.

According to Martins et al. (1999) low plant height is directly related to plant stand and according to Rocha (2009) this is because there is not as much competition between soya beans for light as there would be with larger populations. Schanoski et al. (2011) reported that one of the problems in soya cultivation is harvest losses due to pod insertion height, as this can cause significant losses.

The acceptable level of soya losses per hectare is 60 Kg (AGUILA et al., 2011), which means that of the four cultivars evaluated, only the Coodetec 205 cultivar showed losses greater than 60 Kg ha^{-1} , with losses of 88 Kg ha^{-1} , and this cultivar was the one with the lowest first pod insertion height (6.1 cm) (Table 3).

The losses were not correlated with any of the agronomic characteristics assessed (Table 4). However, Silveira and Conte (2013) mention that harvest losses may be related to low first pod insertion, especially if it is in the first 5 cm above the ground. The researchers also argue that a large part of the losses in harvesting soya beans are related to the number of branches, and the greater the number of branches, the more losses there are. Queiroz et al. (1978) and Silveira and Conte (2013) mention that the number of branches does not influence crop yield, and that a high number of branches is related to low plant stand.

Some correlations were studied in this research and it was found that some variables correlate. Firstly, it was found that the number of pods per plant (VP) was positively correlated with the number of internodes (NE) at 1% probability, in the order of 0.64 (strong). The number of grains per plant (GP) correlated positively at 5% probability with the number of internodes (NE), in the order of 0.60 (strong), and also correlated positively at 1% probability with the number of pods per plant (VP), with 0.98 (very strong). Plant height (PH) was positively correlated at 5% probability with the number of internodes (NE), at 0.57 (medium). The mass of a thousand grains (M1000) was negatively correlated at 5% probability with the height of insertion of the first pod (IV), at -0.59 (average). Yield (PROD) was positively correlated at 1% probability with the number of internodes (NE) and plant height (AP), in the order of 0.76 and 0.68 respectively (both strong) (Table 4).

Table 4: Simple correlation test between the variables (r): number of internodes per plant (NE), number of pods per plant (VP), number of grains per plant (GP), height of insertion of the first pod (IV) in cm, plant height (AP) in cm. As well as the thousand grain mass (M1000) in grams, grain yield per area (PROD) in kg ha^{-1} and grain losses per area (PER) in kg ha^{-1} . Dois Vizinhos, PR, 2014.

	VP	GP	IV	AP	M1000	PROD	PER
NE	0,64**	0,60*	-0.22ns	0,57*	-0.23ns	0,76**	-0,14^{TM}
VP	-	0,98**	-0.35ns	0.01ns	-0.28ns	0.35ns	0,21^{ns}
GP	-	-	-0.45ns	-0.05ns	-0,13^{TM}	0.40ns	0,31^{ns}
IV	-	-	-	0.48ns	-0,59*	-0,17^{TM}	-0,334^{ns}
AP	-	-	-	-	-0.34ns	0,68**	-0,28^{ns}
M1000	-	-	-	-	-	0.19ns	0,40^{ns}
PROD	-	-	-	-	-	-	0,09^{ns}

** significant at the 1% probability level ($p < .01$), * significant at the 5% probability level ($.01 =< p < .05$), ns not significant ($p >= .05$) and $p = 0.0000$ means $p < 0.0001$. NE: number of internodes, VP: pods per plant, GP: grains per plant, IV: insertion of the 1ª pod, AP: plant height, M1000: mass of a thousand grains, PROD: grain yield, PER: grain losses.

The correlation between grains per plant and the number of internodes is positive, i.e. increasing the number of internodes increases the number of grains per plant. Perini et al (2012) also found this correlation when assessing different soya cultivars, in which the correlation was positive at 0.68, similar to the 0.60 found in this experiment (Table 4).

There was a strong correlation between the number of pods per plant and grains per plant, in the order of 0.98, a higher correlation value than that found by Souza et al (2013), which was 0.92. Dalchiavon and Carvalho (2012) also found a positive correlation between these traits (0.885).

Carvalho et al. (2002), through numerous crosses of soya cultivars and subsequent evaluation, mention that the correlation between yield and number of internodes varies greatly from cultivar to cultivar and especially between seasons, both being positive correlations. However, in association with these results, the research carried out by Carvalho et al. (2002) shows that the later the sowing, the greater the correlation between these two parameters, which in this study was 0.76 (Table 4).

Plant height correlates directly with the number of internodes (Table 4), in agreement with Taiz and Zeiger (2004). A greater plant height can promote not only a greater number of internodes, but also greater elongation between internodes (SOUZA et al., 2013), with a positive correlation of 0.57 being observed in this experiment (Table 4).

Rocha (2009), in experiments evaluating soya cultivars, mentions that soya yield is positively correlated with plant height, i.e. the higher the yield, the greater the plant height. In this experiment, the Brasmax Força RR cultivar had the highest yield and at the same time the highest plant height, while the Agro Norte 78017 cultivar had the lowest yield and also the lowest plant height (Table 3), with a correlation of 0.68 between these variables being observed in this study (Table 4).

CONCLUSIONS

Under the conditions of the study, the Brasmax Força RR cultivar is the most productive with a yield of 2,885.5 Kg ha^{-1} , followed by Coodetec 205 (2,325.9 Kg ha^{-1}) and Agroeste 3610 Intacta (1,778.3 Kg ha^{-1}).

Grain loss of 88 kg ha^{-1} was observed for the Coodetec 205 cultivar, which is above the acceptable level.

BIBLIOGRAPHICAL REFERENCES

AGROESTE. Soya cultivars: Agroeste 3610 Intacta RR2 PRO. 2015. Available at <http://www.agroeste.com.br/intactarr2pro/> Accessed on 25 Sep. 2015.

AGUILA, L.S.H.; AGUILA, J.S.; THEISEN, G. Harvest losses in the soya crop. Pelotas: Embrapa Clima Temperado. 1 ed. 12p. 2011. Available at: <http: //ainfo.cnptia. embrapa.br/digital/bitstream/item/79567/1/comunicado - 271.pdf>. Accessed on: 13 December 2015.

BONATO, E.R.; BONATO, Y.; LIDIA, A. Cultivars that made the history of soya in Rio Grande do Sul. Passo Fundo: Embrapa Trigo, 2002, p.15.

BRASMAX. Soya cultivars: Brasmax Força RR. 2014. Available at <http://www.brasmaxgenetica.com.br/producto.php?id=12&r=S> Accessed on 25 Sep. 2015.

CARVALHO, C.G.P.; ARIAS, C.A.A.; TOLEDO, J.F.F.; OLIVEIRA, M.F.; VELLO, N.A. Correlations and path analysis in soya bean lines sown at different times. Pesquisa

Agropecuária Brasileira, Brasília, v. 37, n. 3, p. 311-320, mar. 2002.

CONTINI, E.; TALAMINI, D.J.D.; VIEIRA, P.A. World commodities scenario: chicken, soya and corn. 2015. Available at <http://ainfo.cnptia.embrapa.br/digital/bitstream/item/91011/1/final7198.pdf> Accessed on 26 September 2015.

COODETEC. South region product guide. 2010.

DALCHIAVON, F.C.; CARVALHO, M.P. Linear and spatial correlation of production components and soya productivity. Ciências Agrárias, Londrina, v. 33, n. 2, p. 541-552, abr. 2012.

GARCIA, A.; PÍPOLO, A.E.; LOPES, I.O.N.; PORTUGAL, F.A.F. Soya crop establishment: season, cultivars, spacing and plant population. Londrina: Embrapa Soja. 2007. 12p. Available at: <http://www.infoteca.cnptia.embrapa.br/bitstream/doc/470313/1/circtec51.pdf>. Accessed on: 13 December 2016.

JOHN DEERE. John Deere 1109 VacuMeter seeder technical leaflet™ . 2014. Available at <http://www.deere.com.br/pt_BR/docs/zmags/products/equipment/planters/1100_series/1109.html> Accessed 25 Sep. 2014.

MARCHIORI, L.F.S.; CÂMARA, G.M.S.; PEIXOTO, C.P.; MARTINS, M.C. Vegetative performance of soya cultivars *[Glycine max* (L.) Merrill] in normal and off-season. Scientia *Agricola* vol.56 n.2. 1999.

MARTINS, C.A. Evaluation of Agronomic Characteristics of Soya Strains with or without Lipoxygenases in the Seeds. Master's thesis. Viçosa, 1999.

NEVES, J.A. Agronomic performance of soya genotypes under low latitude conditions in Teresina-PI. 2011. 94 f. Monograph (Postgraduate Diploma in Agronomy). Postgraduate Centre in Agricultural Sciences, Federal University of Piauí, Terezina. 2011, p. 28-32.

NEW HOLLAND (a). TM 7010 Semipowershift tractor technical brochure. 2017. Available at <http://agriculture.newholland.com/br/pt/Products/Tractors/TM7000/Documents

/BRD1589%20-%20TM%20SPS.pdf> Accessed 25 Sep. 2017.

NEW HOLLAND (b). TC 5070 *Hydro* harvester technical leaflet. 2017. Available at <http://agriculture.newholland.com/br/pt/Products/Combine/TC/Documents/TC 5070_TC5090E.pdf> Accessed 25 Sep. 2017.

PERINI, L.J.; JÚNIOR, N.S.F.; DESTRO, D.; PRETE, C.E.C. Yield components in soya cultivars with determinate and indeterminate growth. Ciências Agrárias, Londrina, v. 33, suplemento 1, p. 2531-2544, 2012.

QUEIROZ, E.F.; NEUMAIER, N.; TORRES, E.; TERAZAWA, F.; PALHANO, J.B.; PEREIRA, L.A.G.; BIANCHETTI, A.; YAMASHITA, J. *Technical recommendations for soya harvesting.* Londrina: Embrapa Soja, 1978. 32 p.

RIBEIRO, P.J.; BIAGGIONI, M.A.M. Economic-energy evaluation of combined maize drying. Energia na Agricultura, v.21, p.34-54, 2006.

ROCHA, R.S. Evaluation of soya varieties and strains in low latitude conditions. 2009. 59 p. Dissertation (Master's Degree in Agronomy). Postgraduate Programme in Agronomy, Federal University of Piauí, Terezina, 2009.

SCHANOSKI, R.; RIGHI, E.Z.; WERNER, V. Losses in the mechanised harvesting of soya beans *(Glycine max)* in the municipality of Maripá - PR. *Revista* Brasileira de *Engenharia Agrícola* e *Ambiental* - Agriambi., v.15, n.11, p.1206-1211, 2011.

SEAB. Soya: Analysis of the Agricultural Situation. 2017. Available at <http://www.agricultura.pr.gov.br/arquivos/File/deral/Prognosticos/2017/Soja_2016_17.pdf> Accessed on 27 Aug. 2017.

SILVA, F.A.S.; AZEVEDO, C.A.V. de. Version of the Assistat computer programme for the Windows operating system. Revista Brasileira de Produtos Agroindustriais, Campina Grande, v.4, n.1, p.71-78, 2002.

SILVEIRA, J. M.; CONTE, O. *Determining losses in soya harvesting: as measured by Embrapa.* 1.ed. Londrina: Embrapa Soja, 2013. 29 p.

SMIDERLE, O.J.; GIANLUPPI, V.; GIANLUPPI, D.; JÚNIOR, A.L.M.; ZILLI, J.É.; NECHET, K.L.; BARBOSA, J.F.; MATTIONI, J.A.M. Soya cultivation in the cerrado of Roraima. Boa Vista: Embrapa Roraima, September 2009. Available at:

<http://sistemasdeproducao.cnptia.embrapa.br/FontesHTML/Soja/CultivodeSoja noCerradodeRoraima/lavouras.htm> Accessed on 27 Sep. 2014.

SOUZA, C.A.; FIGUEIREDO, B.P.; COELHO, C.M.M.; CASA, R.T.; SANGOI, L. Plant architecture and soybean productivity resulting from the use of growth reducers. *Bioscience Journal.* Uberlândia, v. 29, n. 3, p. 634-643, May/June 2013.

TAIZ, L.; ZEIGER, E. Trad.: SANTARÉM, E.R. *Plant physiology*. 3.ed. Porto Alegre: Artmed, 2004, 613p.

CHAPTER 6

HUMIC/FULVIC ACIDS AND NITROGEN ON MAIZE PRODUCTIVITY

Paulo Roberto Rabelo
Lucas Link
Eduardo Júnior de Oliveira da Silva
Vanderson Vieira Batista
Karine Fuschter Oligini
Ana Claudia Schllemer dos Santos
Paulo Fernando Adami

Abstract: The aim of this study was to verify the use of a biostimulant based on humic/fulvic acids, associated or not with nitrogen, and to assess its influence on the morphological characteristics and yield components of maize. The study was carried out at the Federal Technological University of Paraná - Dois Vizinhos Campus in a randomised block design with five treatments and four replications. The treatments were: Treatment 1: ^ dose (0.75 L ha^{-1}) of biostimulant at V4 + ^ dose (0.75 L ha^{-1}) at V8 + 170 kg ha^{-1} of N; Treatment 2: recommended dose of biostimulant (1.5 L ha^{-1}) at V4 + 170 Kg ha^{-1} of N; Treatment 3: Recommended dose of the biostimulant (1.5 L ha^{-1}) at V8 + 170 Kg ha^{-1} of N; Treatment 4: Recommended dose of the biostimulant (1.5 L ha^{-1}) at V4; Treatment 5: 170 Kg ha^{-1} of N. Analysis of variance was applied and Tukey's test was applied when significance was found. The morphological characteristics, stalk diameter, first ear insertion height and final plant height, were not influenced by the treatments evaluated. For the yield components, it was observed that the mass of 100 grains was lower when nitrogen was not used (treatment 4), but there was no effect on the other components, rows per ear, grains per row, grains per ear and yield, which was 15,141 Kg ha^{-1} . The morphological characteristics of the plants and the crop's yield components were not influenced by the treatments evaluated, except for grain mass, which was lower when nitrogen was not used.

INTRODUCTION

Maize (Zea Mays) is considered a very old and important agricultural crop for the world, being a strategic product for global food security, which is used in human and animal nutrition (ALVES and AMARAL, 2011). The cereal is the main component of the animal diet of cattle, pigs and poultry (ABIMILHO, 2006), sectors that use around 70% of the maize produced worldwide (PAES, 2006). Castoldi et al. (2011) also point out that the maize plant is the main fodder used in the ensiling process for the whole plant or just the wet grain. This context demonstrates the importance of maize for humanity.

According to Pinotti (2013), maize has high production potential and high nutritional value, which is why it is of great social and economic importance to society. Galvão et al. (2014) point out in their studies that there has been an evolution in the maize production system, in which they report that cereal cultivation has undergone a transition from small-scale, backward agriculture to large-scale, modern agriculture over the last 70 years. However, the researchers point out that producers are currently highly dependent on external inputs, mainly seeds and fertilisers.

It is known that, as a grass, maize is highly responsive to the addition of inputs, especially nitrogen, and grain yield is directly dependent on the efficiency of this nutrient (SANGOI and ALMEIDA, 1994). Several studies have highlighted the importance of nitrogen (N) for maize crops, as this nutrient is directly linked to growth, development and increased grain yield (SANGOI and ALMEIDA, 1994); (FRANCO et al., 2013); (STRIEDER, 2006).

As well as N, other nutrients are essential for successful maize cultivation. Humic and fulvic acids can be used as alternative inputs, as their chemical properties can result in increased productivity due to the benefits that the acids promote for the physical and chemical structure of the soil and for the plant's metabolism (CARON et al., 2015). The researchers emphasise that humic and fulvic acids, in addition to the traditional cultivation system, should be studied, as they can have an influence on crop potential.

Studies have shown a positive effect of humic and fulvic acids on the development of tomato seedlings (BERNARDES et al., 2011), in promoting the growth of pineapple seedlings propagated in vitro (BALDOTTO et al., 2009) and in maize Zandonadi et al., (2007) observed that humic acids promote the induction of lateral roots. However, there are no studies in the literature on the effect of these substances on maize grain yield.

In this context, the aim of this study is to evaluate the use of a biostimulant based on humic/fulvic acids, associated or not with nitrogen, and to see how it affects the morphological characteristics of the plants and the yield components of the maize crop.

MATERIALS AND METHODS

The study was conducted at the Experimental Farm of the Federal Technological University of Paraná - Dois Vizinhos Campus (UTFPR - DV), located in the south-west region of the state of Paraná, with latitude 25°42'S, longitude 53°08'W and an average altitude of 560 metres. The region's climate is classified as Cfa, with hot summers, temperatures above 22°C and more than 30 mm of rain in the driest month (ALVARES et al., 2013). The soil is classified as Latossolo Vermelho Distroférrico (BHERING et al., 2008) and the historical average rainfall in the region is approximately 2,044 mm per year (POSSENTI et al., 2007).

A randomised block design was used, with five treatments and four replications, giving 20 sample units (UAs). The treatments consisted of different ways of using a commercial product based on humic/fulvic acids (biostimulant), associated or not with nitrogen. The treatments evaluated were: Treatment 1: use of ^ dose (0.75 L ha^{-1}) of biostimulant at development stage V4 + ^ dose (0.75 L ha^{-1}) at development stage V8 + 170 kg ha^{-1} of N; Treatment 2: use of the recommended dose of biostimulant (1.5 L ha^{-1}) at development stage V4 + 170 Kg ha^{-1} of N; Treatment 3: use of the recommended dose of the biostimulant (1.5 L ha^{-1}) at development stage V8 + 170 Kg ha^{-1} of N; Treatment 4: use of the recommended dose of the biostimulant (1.5 L ha^{-1}) at development stage V4; Treatment 5: use of 170 Kg ha^{-1} of N.

Each AU consisted of 36 m^2 , 10 metres long and 3.6 metres wide, covering 8 rows of maize crops with 45 cm spacing between rows. The evaluations were carried out in the observation units (OU), which were made up of the two central rows of the AUs, discarding the first and last linear metres of these rows, resulting in 7.2 m of land.[2]

Preparation of the experimental area began 30 days before sowing the maize, with desiccation of the Avena strigosa grown in the area during the winter period. Desiccation was carried out using the chemical herbicide Glyphosate (1.5 L ha^{-1}).

The study was conducted with the 30F53 VYHY hybrid, which was sown on 7 September 2016 in a no-till system, with a density of 80,000 seeds ha^{-1} and spacing of 45 cm between crop rows. The base fertiliser in the sowing furrow was 350 kg ha^{-1} of chemical fertiliser, formulated 16-34-00 (N-P2O5-K2O). In addition, 150 Kg ha^{-1} of KCl was added, supplied by throwing the crop into the furrow shortly after sowing.

The biostimulant was applied using a CO_2 compressor attached to a bar with 4 nozzles.

The source of N used was urea (45% N), which was applied manually, by spreading and top dressing, when the maize plants had four fully developed leaves (vegetative stage V4).

During crop development, weed control was carried out with atrazine (6 L ha^{-1}) at vegetative stage V4. There was no damage to the maize plants in terms of insect attacks or fungal disease problems, so there was no need to apply fungicides or insecticides. However, at VT, 0.4 litres of Abacus was applied as a preventative measure in order to manage any diseases that might occur.During the course of the study, when the maize was at V2, soil was collected from the depths of 0 to 10 cm and 10 to 20 cm, which was sent to the laboratory for analysis, and the summary of the analysis is shown in Table 1.

Table 1. Summary of soil chemical analysis (0 to 10 and 10 to 20 cm), Dois Vizinhos - PR, May (2018).

	MO	K	Ca	Mg	Al3+	P	Ass	Fe	Zn	pH	V
		------ cmol			/dm3$_c$	-------- mg/dm3					
	g/dm^3	---------					--------			CaCl2	(%)
0 - 10	41,55	0,30	4,40	1,60	0,00	8,90	0,00	0,00	0,00	5,10	59,55
10 - 20	32,17	0,13	2,40	1,20	0,35	1,02	0,00	0,00	0,00	4,40	41,08

When the maize kernels were between 20 and 22% moist (05/02/2016), morphological characteristics and yield components were assessed.

As for morphological characteristics, final plant height (cm), ear insertion height (cm) and stalk diameter (cm) were assessed on 10 plants per OU. Both values were obtained using a tape measure, considering the distance between ground level and the main ear insertion internode as the ear insertion height, and the distance between the ground and the apex of the shoot as the final plant height. The diameter of the stalk was obtained between the first and second internode visible above the ground, by circling the stalk with a tape measure, and the value obtained was divided by 3.14 (π). The arithmetic mean of the values obtained from the plants in each OU was calculated and used to analyse the data.

The yield components assessed were number of grains per row, number of rows per ear, number of grains per ear, mass of 100 grains (g) and yield (Kg ha^{-1}).

Ten ears per OU were collected by counting the number of rows per ear and the number of grains in one of the rows. The arithmetic mean of the ears from each OU was calculated and used in the statistical analysis. By multiplying the number of rows per ear and the number of grains per row in the respective OUs, the number of grains per ear was obtained. Yields were analysed by collecting the ears from each OU, which were threshed using a grain whisk attached to a tractor. Impurities were removed and the grain sample obtained was weighed on a precision scale (1 g), the grain moisture was measured using a universal moisture determiner, the yield of the OU was calculated at 13% moisture and extrapolated to hectare (Kg ha^{-1}).

The mass of 100 grains was obtained by sampling one hundred grains of maize at home in each UO, weighing them and correcting the humidity to 13% four times, considering the average value obtained between the four repetitions for the statistical analysis.Analysis of variance (F test) was applied and when there was significance, Tukey's test was applied at 5% probability, using Assistat 7.7beta software (SILVA and AZEVEDO, 2002).

RESULTS AND DISCUSSIONS

Table 2 shows the variables relating to the morphological characteristics of the maize plants, and shows that there was an effect of the treatments studied on.

Average values of 2.37 cm were observed for stalk diameter, with ear insertion height of 133.46 cm and final plant height of 246.06 cm (Table 2). These values are higher than those observed by Batista et al. (2018), who also evaluated various maize hybrids in the municipality of Dois Vizinhos - PR, and found a stalk diameter of 2.09 cm, ear insertion height of 103.8 cm and final plant height of 210.5 cm for the P30F53VYHR hybrid, a difference possibly related to the growing season of the crop.

As for the yield components shown in Table 3, there was an effect of the treatments only for the mass of 100 grains, in which treatment 4, the recommended dose of the biostimulant (1.5 L ha^{-1}) at development stage V4, showed the lowest values (38.49 g), differing from the other treatments. This result demonstrates the importance of nitrogen for grain mass in maize crops.

Table 2. Influence of humic/fulvic acids and nitrogen on stalk diameter, ear insertion height and final plant height in maize. Dois Vizinhos - PR (2018).

Treatments Variables	Stalk diameter (cm)	Ear insertion height (cm)	Final plant height (cm)
Treatment 1	2,41^{NS}	129,83^{NS}	245,83^{NS}
Treatment 2	2,36	134,16	244,50
Treatment 3	2,34	138,66	249,00
Treatment 4	2,37	133,58	246,91
Treatment 5	2,37	131,08	244,08
Average	2,37	133,46	246,06
DMS	0,17	12,31	17,20
CV (%)	3,14	4,09	3,10

NS: Not significant. Treatment 1: ^ dose (0.75 L ha^{-1}) of the technical recommendation of the biostimulant at development stage V4 + ^ dose (0.75 L ha^{-1}) at development stage V8 + 170 Kg ha^{-1} of N (V4); Treatment 2: recommended dose of the biostimulant (1.5 L ha^{-1}) at development stage V4 + 170 Kg ha^{-1} of N (V4); Treatment 3: Recommended dose of the biostimulant (1.5 L ha^{-1}) at development stage V8 + 170 Kg ha^{-1} of N (V4); Treatment 4: Recommended dose of the biostimulant (1.5 L ha^{-1}) at development stage V4; Treatment 5: 170 Kg ha^{-1} of N (V4).

In their studies, Sangoi and Almeida (1994) found that increasing nitrogen levels in maize resulted in a linear increase in the crop's grain mass. Knowing that treatment 4 was the only one without N in its constitution, it is assumed that the biostimulant was unable to meet the plant's demand for nitrogen, resulting in a lower 100-grain mass when N was not used as a top dressing.

Although grain mass was influenced by the treatments, there was no effect on the other yield component variables, number of rows per ear, number of grains per row, number of grains per ear and yield (Table 3). There were average values of 15.89 rows per ear, with 33.10 grains per row, resulting in 530.26 grains per ear (Table 3).

The average yield between the treatments was 15,141.62 kg ha^{-1} (Table 3). When comparing the average yield of the experimental crop with the average yield of maize in Brazil, the southern region of Brazil and Paraná in the same period, which according to CONAB (2018) was 5,556, 7,398 and 8,169 Kg ha^{-1} , respectively, it can be concluded that the experimental crop achieved high levels of yield, a fact possibly linked to the favourable conditions of temperature and water availability observed during the study, the fertility of the soil (Table 1) and the levels of investment in basic fertilisation.

Table 3. Influence of humic/fulvic acids and nitrogen on the number of rows per ear, number of grains per row, number of grains per ear, mass of 100 grains and maize yield. Dois Vizinhos - PR (2018).

Treatments Variables	Row per ear	Grains per row	Grains per ear	Mass 100 grains (g)	Productivity (Kg ha)$^{-1}$
Treatment 1	15.30NS	32.10NS	515.80NS	41,58 a	15.819,10NS
Treatment 2	16,20	34,10	550,70	41,32 a	16.244,03
Treatment 3	15,70	33,25	521,90	40,84 a	14.258,33
Treatment 4	16,00	33,50	536,30	38,49 b	14.008,61
Treatment 5	16,25	32,55	526,62	41,06 a	15.378,05
Average	15,89	33,10	530,26	40,66	15.141,62
DMS	1,81	5,50	82,23	2,01	2.744,67
CV (%)	5,06	7,37	6,88	2,19	8,04

[NS]Not significant. Averages followed by different letters in the column differ statistically by the Tukey test at 5% probability. Treatment 1: ^ dose (0.75 L ha^{-1}) of the technical recommendation of the biostimulant at development stage V4 + ^ dose (0.75 L ha^{-1}) at development stage V8 + 170 Kg ha^{-1} of N (V4); Treatment 2: recommended dose of the biostimulant (1.5 L ha^{-1}) at development stage V4 + 170 Kg ha^{-1} of N (V4); Treatment 3: Recommended dose of the biostimulant (1.5 L ha^{-1}) at development stage V8 + 170 Kg ha^{-1} of N (V4); Treatment 4: Recommended dose of the biostimulant (1.5 L ha^{-1}) at development stage V4; Treatment 5: 170 Kg ha^{-1} of N (V4).

Based on the results observed for the morphological characteristics (Table 2) and the yield components (Table 3), it is surmised that the biostimulant somehow provides an action that meets the plant's demand for N, except for grain mass. However, its use in combination with N does not add to the components evaluated. Despite the positive evidence regarding the use of biostimulants in maize cultivation, further studies need to be carried out to confirm this.

CONCLUSION

The morphological characteristics of the maize plants and the crop's yield components were not influenced by the treatments evaluated, except for grain mass, which was lower when nitrogen was not used.

BIBLIOGRAPHICAL REFERENCES

ABIMILHO. Brazilian Maize Industry Association. Maize guide, technology from the field to the table. 16 f. July 2006.

ALVARES, C.A.; STAPE, J.L.; SENTELHAS, P.C.; MORAES, J.L.G; SPAROVEK, G. Kõppen's climate classification map for Brazil. Meteorologische Zeitschrift, Stuttgart, v. 22, p.711-728, 2013.

ALVES, H.R.A.; AMARAL, R.F. Technical Office for Economic Studies in the Northeast - ETENE. Environment for Studies, Research and Evaluation - AEPA. Coordination of Rural and Agro-Industrial Studies - COERG. Maize production, harvested area and productivity in the Northeast. ETENE Rural Report. N° 16. September 2011.

BALDOTTO, L.E.B, BALDOTTO, M.A., GIRO, V.B., CANELLAS, L.P., OLIVARES, F.L., SMITH, R.B. Performance of 'Vitória' pineapple in response to the application of humic acids during acclimatisation. Revista Brasileira de Ciência do Solo, v. 33, n. 4, 2009.

BATISTA, V. V; LINK, L; GIARETTA, R; SILVA, J. S; ADAMI, P. F. Yield components and productivity of maize hybrids grown in off-season. Pesquisa Aplicada & Agrotecnologia, Guarapuava-PR, v.11, n.2, p.67-75, may- aug., 2018. DOI: 10.5935/PAeT.V11.N2.07

BERNARDES, J.M., REIS, J.M.R., RODRIGUES, J.F. Effect of humic substance application on tomato seedlings. Global Science and Technology, v. 4, n. 3, 2011.

BHERING, S.B., SANTOS, H.G., BOGNOLA, I.A., CÚRCIO, G.R., CARVALHO JUNIOR, W.D., CHAGAS, C.D.S., SILVA, J.D.S. *In:* Embrapa Florestas-Article in conference proceedings (ALICE). *In:* BRAZILIAN CONGRESS OF SOIL SCIENCE, 32nd, 2009, Fortaleza. Soil and bioenergy production: perspectives and challenges: proceedings [Viçosa, MG]: SBCS; Fortaleza: UFC, 2009., 2008.

CARON, V.C.; GRAÇAS, J.P.; CASTRO, P.R.C. Soil conditioners: humic and fulvic acids. Piracicaba: ESALQ/USP, 2015.

CASTOLDI, G.; COSTA, M.S.S.M.; COSTA, L.A.M.; PIVETTA, L.A.; STEINER, F. Cropping systems and the use of different fertilisers in the production of maize silage and grain. Acta Scientiarum Agronomy, v. 33, n. 1, 2011.

CONAB. National Supply Company. Monitoring the 2017/18 Brazilian grain harvest - 4th Conab survey. Brasília - DF. p. 132, v. 5; January 2018.

FRANCO, A.A.N.; MARQUES, O.J.; FILHO, P.S.V. Safrinha maize production systems in

Paraná. National Maize Seminar: Stability and Productivity, v. 12, 2013.

GALVÃO, J.C.C.; MIRANDA, G.V.; TROGELLO, E.; FRITSCHE-NETO, R. Seven decades of evolution of the maize production system. Revista Ceres, v. 61, n. 7, 2014. DOI: http://dx.doi.org/10.1590/0034-737x201461000007

PAES, M.C.D. Technical Circular. Physical, Chemical and Technological Aspects of Maize Grain. Sete Lagoas - MG; Embrapa milho e sorgo, Circ. n. 75, p. 6, ISSN: 1679-1150, December 2006.

PINOTTI, E.B. Evaluation of maize cultivars according to plant population and sowing time. 2013. 134 f. Thesis (PhD in agronomy) - Universidade Estadual Paulista "Júlio Mesquita Filho" Faculdade de Ciências Agronómica. Botucatu, 2013.

POSSENTI, J. C., GOUVEA, A., MARTIN, T. N., CADORE, D. Distribution of rainfall in Dois Vizinhos, Paraná, Brazil. In: Seminar: Agricultural Production Systems-Agricultural, Animal and Forestry Sciences, 140-142. 2007.

SANGOI, L., ALMEIDA, M. L. Doses and times of nitrogen application for maize in a soil with high organic matter content. Pesquisa Agropecuária Brasileira, v. 29, n. 1, p. 01-91, 1994.

SILVA, F.A.S.; AZEVEDO, C.A.V. Version of the Assistat computer programme for the Windows operating system. Revista Brasileira de Produtos Agroindustriais, Campina Grande, v.4, n.1, p.71-78, 2002.

STRIEDER, M.L. Maize responses to reduced row spacing in different management systems. 2016. 100 f. Dissertation (Master's in Plant Science) - School of Agronomy - Federal University of Rio Grande do Sul, Porto Alegre, 2006.

ZANDONADI, D.B.; CANELLAS, L.P.; FAÇANHA, A.R. Indolacetic and humic acids induce lateral root development through a concerted plasmalemma and tonoplast H+ pumps activation. Planta, Berlin, v. 225, p. 1583-1595, 2007.

CHAPTER 7

YIELD COMPONENTS OF TWO MAIZE HYBRIDS GROWN IN "OFF-SEASON" IN SOUTH-WESTERN PARANÁ

Vanderson Vieira Batista
Karine Fuschter Oligini
Ana Claudia Schllemer dos Santos
Eduardo Júnior de Oliveira da Silva
Lucas Link Cleverson Luiz Giacomel
Paulo Fernando Adami

Summary: Maize production in Brazil is divided into two seasons: harvest, with cultivation taking place in the spring-summer period, and off-season in the autumn-winter period. However, it is known that there are a number of factors that can interfere with the productive potential of crops, such as sowing density. The aim of this study was to assess the yield and other agronomic characteristics of two maize hybrids grown in the off-season in the municipality of Dois Vizinhos, Paraná. The study was conducted in the municipality of Dois Vizinhos, Paraná in a randomised block design, with four treatments being evaluated: treatment 1: 2B587Hx maize with a density of 55,000 plants per hectare; treatment 2: 2B587Hx maize with a density of 65,000 plants per hectare; treatment 3: AG9030 maize with a density of 55,000 plants per hectare; treatment 4: AG9030 maize with a density of 65,000 plants per hectare. Plant height, ear insertion height, basal diameter of the stalk, rows per ear, grains per row and mass of a thousand grains showed statistical differences for the treatments studied, but moisture and yield were not influenced by the treatments studied. The productivity of both hybrids was satisfactory, being higher than the national and Paraná averages.

INTRODUCTION

Maize is widely produced throughout the world and is the most produced and consumed cereal worldwide. It is also classified as the second most traded commodity on the market, second only to soya (CONAB, 2016).

Brazil is the third largest producer of maize (Zea mays), behind only the United States and China (USDA, 2015). Maize production in Brazil is divided into two crops: summer maize, which is grown during the spring/summer period and sown between August and November (FORNASIERI FILHO, 2007), and the winter crop, the second crop or "safrinha" as it is popularly known, which is grown during the autumn/winter period and sown between January and February.

In order to grow grain successfully, producers must take into account various factors, such as the choice of genetic material, the plant density to be used and fertiliser levels. According to Argenta et al. (2001), manipulating the plant arrangement is among the most important practices for maximising grain yield. Fornasieri Filho (2007) emphasises that the choice of quality genetic material is also fundamental for successful grain cultivation.

According to data from SEAB/DERAL (2014), the south-western region of Paraná has the largest second-crop maize crop in the state. Sangoi and Silva (2006), on the other hand, report that studies dealing with plant density in the cultivation of off-season maize still lack technical information in order to optimise its performance and viability. The aim of this study was to evaluate two maize hybrids that are recommended and grown in the off-season in the municipality of Dois Vizinhos - PR, checking the yield components of the materials in two plant populations.

MATERIALS AND METHODS

The experimental field was located in the community of São Valentim, in the rural area of the municipality of Dois Vizinhos, Paraná. It has an altitude of 530 metres, flat topography and soil classified by Bhering et al. (2008) as Latossolo Vermelho distrófico. The climate is classified according to Alvares et al. (2013) as Cfa (humid subtropical climate without a defined dry season with an average temperature of the warmest month of 22°C and infrequent frosts). Rainfall in 2014 was approximately 2,000 mm, which is distributed throughout the year (IAPAR, 2014).

During the experiment, data was collected on maximum temperature, minimum temperature, average temperature and rainfall, which are shown in Figure 01.

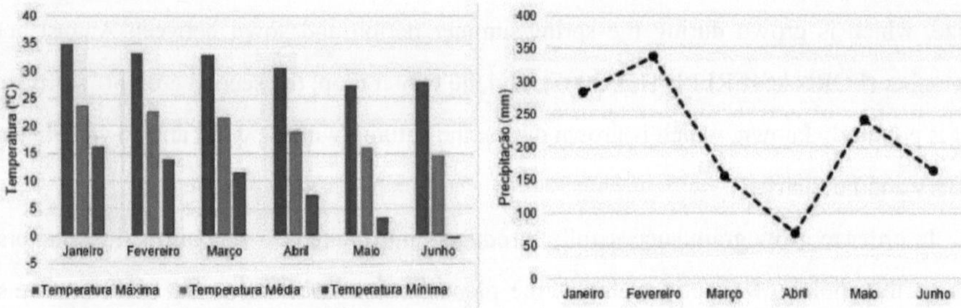

Figure 01: Precipitation (mm) and maximum temperature (°C), minimum temperature (°C) and average temperature (°C) recorded during the safrinha maize growing season (2015). Source: GEBIOMET, 2016.

Soil was also collected at a depth of 0 to 20 cm and sent to the laboratory for chemical composition analyses (Table 01).

Table 01: Summary of soil chemical analysis, (0 - 20 cm), Dois Vizinhos - PR. January 2015.

MO	K	Ca	Mg	Al3+	P	Ass	Fe	Zn	pH	V
g/dm^3	Cmolc/dm3 ---------------				-------- mg/dm3 -----------				CaCl2	(%)
53,2	0,86	11	3,31	0	39,6	9,19	23,7	14,5	5,2	72,5

The study used two maize hybrids, 2B587Hx and AG9030. The 2B587Hx hybrid is of the simple type, has an early cycle (approximately 155 days) and is recommended for off-season planting with a population of 55,000 plants per hectare (DOW SEMENTES, 2013). Maize AG9030 has a super-early cycle, with a technical recommendation of 55,000 to 65,000 plants per hectare in the off-season (SEMENTES AGROCERES, 2016).

The experiment was conducted in a randomised block design with four treatments and three replications, generating 16 experimental units. Each treatment was made up of a hybrid and a cultivation density, thus generating the following combinations: treatment 1: 2B587Hx maize with a density of 55,000 plants per hectare; treatment 2: 2B587Hx maize with a density of 65,000 plants per hectare; treatment 3: AG9030 maize with a density of 55,000 plants per hectare; treatment 4: AG9030 maize with a density of 65,000 plants per hectare.

Each experimental unit consisted of four planting rows, each twelve metres long. For the evaluations, only plants located in the two central rows were used, disregarding the side rows and also the plants in the central row that were one metre from the edge of the planting row, thus generating a useful area of 9 m^2 for each experimental unit.

The maize was sown on 15 January 2015, just after the soya harvest, using a seeder-advocate, which travelled at a speed of 4 km.hr. Solid lubricant (3 g kg^{-1} of powdered graphite) was added at the time of sowing to improve the performance and uniformity of seed deposition. Also added to the planting line was 246 kg ha^{-1} of 2-28-20 chemical fertiliser.

Weeds and volunteer soya beans that appeared were controlled with Atrazine (4 L ha^{-1}). Subsequently, no pest, invasive plant or disease control was carried out, as there was no evidence of their presence.

When the plants were at the V4 phenological stage, i.e. when they had four fully developed leaves, 150kg of nitrogen (urea with 46% N) per hectare was applied manually.

The data was collected when the maize grains had reached a moisture content of approximately 22%, a period when the producer harvests the off-season maize crops in the region.

- Plant height (PA) (cm): the plant height value was obtained using a tape measure, taking into account the distance from the ground to the insertion of the pendant of five plants per experimental unit, then the arithmetic mean was calculated between the values found for each treatment.

- Ear insertion height (AIE) (cm): arithmetic mean of five plants per experimental unit, with the value considered to be between the base of the soil and the insertion point of the first ear.

- Stalk diameter (SD): the diameter of the stalk between the second and third internode of five plants was obtained using a digital caliper, followed by the arithmetic mean of the values found.

- Rows per ear (FE): five ears were collected at random per experimental unit and then the number of rows of grains present in each ear was counted, with the average value of the data for each experimental unit being taken into account for statistical analysis.

- Grains per row (GF): in five ears per experimental unit, the number of grains present in one of the rows was counted and then the arithmetic mean between the values was calculated.

- Productivity (PROD) (Kg ha^{-1}): all the ears present in each experimental unit were collected, the grains were threshed and weighed, the humidity was corrected to 13% and then the production value was calculated in kilograms per hectare.

- Grain moisture (UMD) (%): a 60-gram sample of grains from each experimental unit was weighed and taken to a universal moisture determiner to check the percentage of grain moisture. The average value for each experimental unit was used for statistical analysis.

- Thousand grain mass (MMG) (g): 100 grains were collected from each experimental unit, weighed and then multiplied by a correction factor of 10.

The data was analysed by variance and the means were compared using Duncan's test at a 5% probability level using Assistat software, version 7.7 beta (SILVA e AZEVEDO, 2002).

RESULTS AND DISCUSSION

Plant height and ear insertion height are agronomic characteristics that are usually correlated for maize crops. Table 02 shows statistical differences for the variable plant height (PH), where the treatments with 2B587 maize were taller than the treatments containing AG9030 maize. There were also differences between the treatments established with AG9030 (Table 02), in which the density of 65,000 plants obtained a higher PH than the density of 55,000 plants.

Table 02: Test of means for the following variables: plant height (AP) (cm), ear insertion height (AIE) (cm), basal stem diameter (DBC), moisture content (UMD) (%), rows per ear (FE), grains per row (GF), thousand grain mass (MMG) (g) and productivity (PROD) ($kg.ha^{-1}$) of two maize hybrids AG9030 and 2B587 grown at two densities in the off-season (2015) in the municipality of Dois Vizinhos - Paraná.

Treatments	AP	IEA	DC	UMD
AG9030 - 55	212,00 c	100,07 b	2,87 a	24,67 a
AG9030 - 65	222,33 b	109.33 ab	2,68 b	24,83 a
2B587 - 55	244,67 a	114,13 a	3,05 a	27,17 a
2B587 - 65	243,60 a	117,60 a	2,89 a	25,33 a
F	45,0582**	5,2874*	7,2322*	2,1515 ^{ns}
CV (%)	1,81%	5,20%	3,32%	5,31%
Treatments	**FE**	**GF**	**MMG**	**PROD**
AG9030 - 55	17.60 ab	28,53 b	370,20 a	9587,03 a
AG9030 - 65	18,53 a	28,53 b	361,47 a	11577,78 a
2B587 - 55	17.60 ab	33,27 a	355.29 ab	10153,71 a
2B587 - 65	16,26 b	32.20 ab	345,86 b	9847,20 a
F	4,4470*	3,5704*	5,0187*	2,1985 ^{ns}
CV (%)	4,38%	7,37%	2,20%	10,08%

*, ** and^{ns} : significant at 1%, 5% and not significant respectively.
Averages followed by different letters in the columns differ by Duncan's test at 5% probability.

Statistical differences were also observed for the variable ear insertion height (AIE) (Table 02), in which the treatments with maize hybrid 2B587 obtained a higher AIE when compared to the treatment using maize AG9030 at a density of 55,000 plants. The combination of AG9030 maize + 65,000 plants did not differ statistically from the other treatments (Table 02). Campos et al. (2010) studied forty-nine commercial maize cultivars in the off-season and found no differences in plant height and ear insertion height.

The combination of the AG9030 hybrid + density of 65,000 plants gave the smallest stalk diameter (SD), differing statistically from the rest of the combinations (Table 02). For the other treatments, there were no differences in CD (Table 02). Calonego et al. (2011) and Pinotti (2013) observed in their studies that increasing the plant population reduces the stalk diameter of maize. The manipulation of the plant arrangement alters the quality of the light received by the leaves, so at high densities morphological changes can occur in the architecture and development of the plants, including in the diameter of the stalk (STRIEDER, 2006).

Although the hybrids have different cycles, AG9030 super early (SEMENTES AGROCERES, 2016) and 5B587 early (DOW SEMENTES, 2013), there were no statistical differences between the treatments for the moisture (UMD) variable (Table 02). However, Table 02 shows a slight trend in which the treatments with AG9030 have lower grain moisture contents, 24.67 for the density of 55,000 and 24.83 for the density of 65,000 plants. This slight difference is attributable to the crop's cycle, since AG9030 maize has a slightly shorter cycle.

The variable row per ear (FE) was statistically superior in the AG9030 + density of 65,000 plants combination (18.53) when compared to the 2B587 + 65,000 plants treatment (16.26) (table 02). No statistical differences were observed for the other treatments (table 02). Pinotti (2013) studied two maize hybrids DKB 393 and DKB 330 and found average FE values of 16.3 and 15.5 respectively, similar to the lower values found in this study. He also observed that the cultivar DKB 393 showed a slight reduction in the number of rows on average as the plant population increased, a fact not observed in this study with the hybrids 2B587 and AG9030.

In order to obtain the best yields, it is important that the ears of maize have the highest number of grains per row possible, as this, together with the number of rows, mass of a thousand grains and population density, are the factors responsible for the crop's productivity. The number of grains per row (GF) was statistically superior in the combination that used hybrid 2B587 + a density of 55,000 plants, compared to the two treatments that used AG9030 maize (Table 02). However, the treatment with maize 2B587 + density of 65,000 plants did not differ statistically from the others (Table 02). It can be seen that growing the 2B587 hybrid at the recommended density (55,000 plants ha^{-1} (DOW SEMENTES, 2013)) showed higher GF values when compared to the AG9030 maize treatments.

Table 02 shows that the mass of a thousand grains (MMG) is statistically superior in the AG9030 + density of 55,000 plants and AG3090 + density of 65,000 plants treatments, when compared to the treatment composed of the maize hybrid 2B587 + density of 65,000 plants. The combination of 2B587 + density of 55,000 plants did not differ statistically from the other treatments (Table 02). In studies carried out by Oliveira (2014), there was a linear reduction in MMG as the density increased. Calonego et al. (2011) found no differences when assessing densities of 45, 60 and 75 thousand plants ha^{-1} .

Yield (PROD) was not statistically affected by the combination of treatments, i.e. there were no statistical differences in PROD between the treatments studied (Table 02). These results differ from those of Cruz et al. (2014), who reported that maize is very sensitive to increases in sowing density, which affects the number of ears and their size, thus potentially affecting yield. Sangoi (2000) also reports that at high densities, maize plants tend to increase inter-plant competition for light, water and nutrients, which can affect yields.

As can be seen in Table 01, the fertility of the soil in the area where the experiment was set up had excellent levels of organic matter (53.2 g/dm^3), phosphorus (39.6 mg/dm^3) and potassium (0.86 $Cmolc/dm^3$). Also, with the addition of 246 Kg ha^{-1} of chemical fertiliser in the 2-28-20 formulation when sowing the maize and 150 Kg ha^{-1} of nitrogen at the start of plant development, it can be said that the soil was in optimum condition for growing maize, thus masking the possible effects of intraspecific competition in the higher densities, i.e. providing good plant development and consequently good productivity in all the treatments studied.

IAC (2014) carried out a study in the Paranapanema Valley region of São Paulo in 2010, evaluating various maize varieties grown in the off-season, including AG9030, which obtained a yield of 6,351 kg ha^{-1} . Piana et al. (2008) studied the 2B587 hybrid grown in the municipality of Eldorado do Sul (RS), in the 2006/2007 harvest, and the maize obtained a yield of over 12 tonnes ha^{-1} . Silva et al. (2016) also studied maize 2B587 in Minas Gerais in the 2014/2015 agricultural year, which was subjected to different levels of nitrogen fertilisation and obtained yields of 9.5 t ha^{-1} with 0 nitrogen (N) and just over 11 t ha^{-1} with 240 kg of N.

Another study carried out by IAPAR, which assessed different maize materials grown in twelve municipalities in the state of Paraná, found that maize yields in locations with altitudes below 700 metres were 9,974 Kg ha^{-1} for genetically modified materials and 8,730 Kg ha^{-1} for conventional maize (SHIOGA et al., 2015). In Pato Branco, a city in the south-west of Paraná, the average yield was 13,439 Kg ha^{-1} for genetically modified material and 11,404 Kg ha^{-1} for conventional (SHIOGA et al., 2015).

Considering the production values found in the studies mentioned above, it can be said that the hybrids studied had lower yields than those recorded in other cities in the region, but they are satisfactory when compared to the state's yields obtained below an altitude of 700 metres, since Dois Vizinhos, Paraná, is located in a region with an altitude of 530 metres, and average yields of 10.5 t ha^{-1} and 10 t ha^{-1} were observed for the AG9030 and 2B587 corn, respectively.

CONCLUSIONS

The variables plant height, ear insertion height, basal diameter of the stalk, rows per ear, grains per row and mass of a thousand grains showed differences for the combination of treatments studied.

Moisture and yield showed no statistical differences for the combination of treatments evaluated.

The yields obtained from the AG9030 and 2B587 hybrids were lower than others recorded in the region, but satisfactory for crops grown below an altitude of 700 metres.

BIBLIOGRAPHICAL REFERENCES

ALVARES, C. A.; STAPE, J. L.; SENTELHAS, P. C.; MORAES, J. L. G; SPAROVEK, G. Kõppen's climate classification map for Brazil. Meteorologische Zeitschrift, Stuttgart, v. 22, p.711-728, 2013.

ARGENTA G.; SILVA P.R.F. da; SANGOI L. Plant arrangement in maize: state-of-the-art analysis. Ciência Rural Magazine, v. 31, n. 6, 2001.

BHERING, S. B., SANTOS, H. G., BOGNOLA, I. A., CÚRCIO, G.R., CARVALHO JUNIOR, W. D., CHAGAS, C. D. S., SILVA, J. D. S. In: Embrapa Florestas- Article in conference proceedings (ALICE). In: BRAZILIAN CONGRESS OF SOIL SCIENCE, 32nd, 2009, Fortaleza. Soil and bioenergy production: perspectives and challenges: proceedings [Viçosa, MG]: SBCS; Fortaleza: UFC, 2009., 2008.

CALONEGO J.C.; POLETO L.C.; DOMINGUES F.N.; TIRITAN C.S. Maize productivity and growth in different plant arrangements. Agrarum Magazine. Dourados, v. 4, n. 12, p. 84-90, 2011.

CAMPOS M.C.C.; SILVA V.A.; CAVALCANTE I.H.L.; BECKMANN M.Z.; Productivity and agronomic characteristics of safrinha maize cultivars under no-tillage in the state of Goiás. Ver. Acad., Ciênc. Agrár. Ambient., Curitiba, v.8, n. 1, p. 77-84, jan./mar. 2010.

CONAB. National Supply Company. Brazilian Grain Harvest 2013/14 - 12th Conab Survey. DEAGRO Newsletter (FIESP Agribusiness Department) Electronic version. Available at: <http://www.fiesp.com.br/indices-pesquisas-e-publicacoes/safra-mundial-de-milho-2/>Accessed on: 22/01/2016.

CRUZ J.C.; PEREIRA F.T.F.; PEREIRA FILHO I.A.; OLIVEIRA A.C.; MAGALHÃES P.C. Responses of Maize Cultivars to Variation in Spacing and Density. XXV National Maize and Sorghum Congress. Cuiabá - MG. 2004.

DOW SEEDS. Seed guide 2013. Dow AgroSciences. 2013.

FORNASIERI FILHO, D. Manual da cultura do milho. Jabuticabal. Funesp. 2007

GEBIOMET. Biometeorology study group. Available at: <www.gebiomet.com.br/dowloads.php> Accessed on: 13 April 2016.

IAC. Agronomic Institute. Regional Evaluation of Safrinha Maize Cultivars in the State of

São Paulo. Results 2014. APTA - Middle Paranapanema Regional Centre. Assis (SP), 23/10/2014.

IAPAR. Agronomic Institute of Paraná. Historical averages at IAPAR stations. Available at: <htpp://www.iapar.br/arquivos/Imagens/monitoramento/Medias_Historicas/Francisco_Bltrao.htm>. Accessed on: 27 October 2014.

OLIVEIRA A.G. Desiccation times of ryegrass and ryegrass plus vetch, grazed or with ground cover, on maize performance. 2014. 176 f. Dissertation (Postgraduate Programme in Agronomy) - Federal Technological University of Paraná - UTFPR, Pato Branco, 2014.

PIANA T.A.; SILVA P.R.F.; BREDEMEIER C.; SANGOI L.; VIEIRA V.M.; SERPA M.S.; JANDREY D.B.; Plant density of hybrid maize in early sowing in Rio Grande do Sul. Ciência Rural, Santa Maria, v.38, n.9, p2608-2612, dec. 2008.

PINOTTI E.B. Evaluation of maize cultivars according to plant population and sowing time. 2013. 134 f. Thesis (PhD in agronomy) - Universidade Estadual Paulista "Júlio Mesquita Filho" Faculdade de Ciências Agronómica. Botucatu, 2013.

SANGOI L.; SILVA P.R.F.; Density and population arrangement in maize. 2006. Article in Hypertext . Available at : <http://www.infobibos.com/Artigos/2006_2/Densidade/Index.htm>. Accessed on: 22/12/2015.

SANGOI L; ENDER M.; GUIDOLIN A.F., et al. Evolution of disease resistance of maize hybrids from different seasons in three plant populations. In: BARBOSA, Tânia G. Revista Ciência Rural, Santa Maria, v. 30, n. 1, p. 17-21, 2000.

SEAB/DERAL. Secretariat of Agriculture and Supply of the State of Paraná. Department of Rural Economy. Agricultural production table by municipality. Electronic version . Available at: <http://www.agricultura.pr.gov.br/modules/conteudo/conteudo.php?conteudo=1 37#>. Accessed on: 25 October 2014.

AGROCERES SEEDS. Agroceres seed technical guide. Available at: <http://www.sementesagroceres.com.br/pages/Produto.aspx?p=AG_9030> Accessed on: 30 Dec. 2016.

SHIOGA P.S.; GEREGE A.C.; ARAÚJO P.M.; BIANCO R.; CUSTÓDIO A.A.P.; State Evaluation of Maize Cultivars 2014/2015 Harvest. IAPAR. Technical Bulletin N°85. August 2015.

SILVA, F. A. S.; AZEVEDO, C. A. V. de. Version of the Assistat computer programme for the Windows operating system. Revista Brasileira de Produtos Agroindustriais, Campina Grande, v.4, n.1, p.71-78, 2002.

SILVA R.M.A.; OLIVEIRA E.; SILVA A.V.; OLIVEIRA JUNIOR A.C.; TEIXEIRA G.V.; BREGAGNOLI M.; Productivity of conventional maize 2B587 in different doses of Nitrogen. Available at: <http://webcache.googleusercontent.com/search?q=cache:http://www.abms.org.br/cnms2016_trabalhos/docs/970.pdf&gws_rd=cr&ei=5MWAWPbSMMqawATv4r2oCg> Accessed on: 18 January 2017.

STRIEDER M.L. Maize responses to reduced row spacing in different management systems. 2016. 100 f. Dissertation (Master's in Plant Science) - School of Agronomy - Federal University of Rio Grande do Sul, Porto Alegre, 2006.

USDA. United States Department of Agriculture. World Corn Harvest 2014/15 - 12th USDA Survey. DEAGRO Newsletter (FIESP Agribusiness Department) Electronic version, April 2015. Available at: <http://www.fiesp.com.br/indices- pesquisas-e-publicacoes/safra-mundial-de-milho-2/>Accessed on: 22/01/2016.

CHAPTER 8

FUNGAL MICROBIOTA IN SAFRINHA MAIZE SEEDS

Vanderson Vieira Batista
Cleverson Luiz Giacomel
Eduardo Júnior de Oliveira da Silva
Karine Fuschter Oligini
Ana Claudia Schllemer dos Santos
Marciani Balbinotti France
Paulo Fernando Adami

Abstract: Safrinha corn in Paraná is harvested in autumn, a favourable period for the development of fungal microbiota. The aim of this study was to assess the effects of plant density and nitrogen fertiliser levels on the health of maize ears and grains, observing the presence of fungal microbiota, in off-season maize grown in the municipality of Dois Vizinhos, Paraná. The experiment was conducted in a randomised block design, in subdivided plots, with three replications. Three factors were analysed: Factor A, sowing density (45,000, 55,000, 65,000 and 75,000 plants ha^{-1}), in the main plots; Factor B, nitrogen fertiliser levels (0, 50, 100 and 150 kg of N ha^{-1}), in the subplots; and Factor C, two harvest times: season 1, when the maize had reached physiological maturity, and season 2, when the producer normally harvests off-season maize (grain moisture of around 25%). The percentage of sprouted grains on the cob, the presence of fungi on the cobs (PFE) and the presence of fungi on the grains were evaluated. There was an interaction between density and nitrogen fertiliser levels for the number of sprouted grains on the cob. The length of time the maize remains in the field harms grain health, increasing the number of sprouted grains on the cob and the incidence of fungal microbiota of the genera Aspergillus spp., Fusarium spp. and Penicillium spp.

INTRODUCTION

Maize (Zea mays L.) is widely grown around the world, with Brazil being the third largest producer in the 2013/2014 agricultural harvest, with 79 million tonnes FIESP (2014). Among the Brazilian states, Paraná stands out in terms of production, which is divided into two seasons: harvest corn, sown between August and September Fornasieri Filho (2007), and off-season corn, sown in January and February.

However, low temperatures, associated with high relative humidity and short days, combined with increased rainfall in the autumn period (harvest time for off-season maize) can increase the incidence of stalk rot, the spread of pathogens and plant lodging, Fontoura et al. (2006). These factors can also lead to problems with burnt grains, which are common in off-season crops, as a result of ear rot, caused mainly by fungi that are present in the field (PINTO, 2005).

Therefore, the quality of the grain can be compromised by fungal infection, as these microorganisms can produce mycotoxins, which have the potential to cause damage to human and animal health (FARIAS et al., 2000). Fusarium spp. and Aspergillus spp. stand out among the main fungi found on maize. These fungi cause burnt grains and have toxigenic potential, being highly harmful to human and animal health, producing substances called mycotoxins. A diet of cattle, poultry or pigs made up of grains with high levels of mycotoxins means danger and losses, as they can cause irreversible damage to the animals' health, jeopardising the integrity of meat, milk and products derived from intoxicated animals (PINTO, 2005).

Knowing that microorganisms can be harmful to animal diets and that they can come from crops, the aim of this study was to assess the impact that plant density and nitrogen fertiliser levels have on the health of corn cobs and grains, observing the presence of fungal microbiota in off-season corn grown in the municipality of Dois Vizinhos, Paraná.

MATERIALS AND METHODS

The experiment was conducted in the municipality of Dois Vizinhos, Paraná, latitude 25° 48'09" South, longitude 53°06'28" West and an altitude of 530 metres. The experimental area has flat topography and the soil is classified as Latossolo Vermelho Distrófico (BHERING et al., 2008). The climate is classified as Cfa - humid subtropical climate without a defined dry season with an average temperature of the warmest month of 22°C and infrequent frosts (ALVARES et al., 2013). Rainfall is approximately 2,000 mm in 2013, which is distributed throughout the year IAPAR (2014).

The maximum, minimum and average temperature and rainfall data recorded during the experiment are shown in Figure 01.

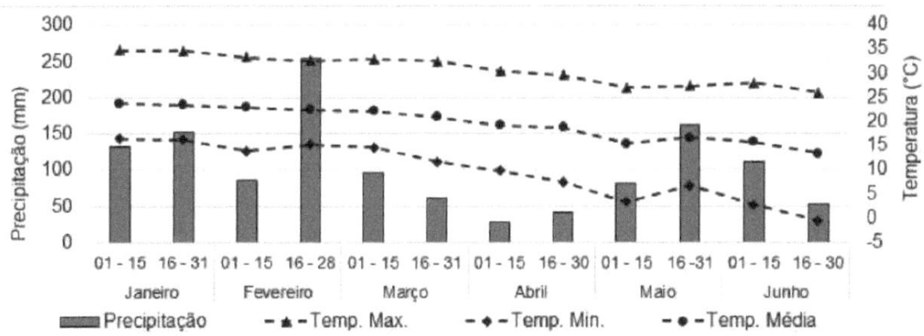

Figure 1. Precipitation (mm) and maximum (Max), minimum (Min) and average (°C) temperatures recorded during the safrinha maize growing season (2015) GEBIOMET (2016).

The experiment was conducted in a randomised block design, in subdivided plots, with three replications, containing a total of 48 experimental units (EU). Three factors were analysed: Factor A, sowing density (45,000, 55,000, 65,000 and 75,000 plants ha^{-1}), in the main plots; Factor B, nitrogen fertiliser levels (0, 50, 100 and 150 kg of N ha^{-1}), in the sub-plots; and Factor C, two harvest times: season 1, when the maize had reached physiological maturity, and season 2, when the producer normally harvests the maize off-season (grain moisture of around 25%).

Each EU consisted of four planting rows twelve metres long. To carry out the evaluations, the ears of the plants from the two central rows of the sub-plot were used, with the side rows being disregarded. The plants from the central row that were one metre from the edge of the planting row at each end were also discarded from the evaluations, thus generating a useful area of 9 m^2 for each EU.

Hybrid maize 2B587Hx was used, which is classified as an early-cycle simple hybrid. It has the characteristics of low-growing plants with high physical resistance, good rooting and stuffing, easy threshing and grain with a yellow-orange colour. Recommended for grain production, it has high production potential, good ear health and drought tolerance. It is recommended for off-season planting with a cycle of approximately 155 days and populations ranging from 50,000 to 55,000 plants per hectare (DOW SEMENTES, 2013).

The maize was sown on 15 January 2015, shortly after the soya harvest, using a John Deere hydraulic drag seeder coupled to a New Holland TL-95E tractor, which travelled at a speed of 4 km.hr. Solid lubricant (3 g kg^{-1} of powdered graphite) was added at the time of sowing in order to improve the performance and uniformity of seed deposition. Also added to the planting line was 246 kg ha^{-1} of 2-28-20 chemical fertiliser.

Weeds that appeared and volunteer soya beans were controlled by applying Atrazine (4 L ha^{-1}) when the maize plants were at V3. Subsequently, no pest, invasive plant or disease control was carried out, as there was no evidence of their presence. The source of nitrogen (N) used was urea with 46% N. The nitrogen fertiliser levels were applied sixteen days after sowing, on 31/01/2015, as a top dressing and manually, in a single dose, when the plants were at phenological stage V4, i.e. when they had four fully developed leaves. The urea was applied in conditions of high soil humidity, with 96 mm of rain having fallen the night before the application.

The ears for the evaluations were collected manually and randomly, evaluating five ears per EU. The ears reached physiological maturity on 26/05/2015, the harvest date for season 01. The second season was evaluated on 16/06/2015.

Evaluations were carried out to check the percentage of:

- Sprouted grains in the ear (GBE) (%): observing whether the ears contained sprouted grains. The number of ears was recorded and then the value was transformed into a percentage.

- Presence of fungi on the ears (PFE) (%): carried out visually on each ear to check for the presence of fungi. The value was transformed into a percentage.

- Presence of fungi in the grains (PFG) (%): evaluations of fungal microbiota in the grains were carried out using a health test, using the incubation method on filter paper (Blotter Test) developed by Neegaard and modified with water restriction by Machado et al. (2003). The five ears were threshed and their grains mixed to homogenise the sample. Next, 100 grains were chosen at random and subjected to surface disinfestation using 2% sodium hypochlorite for two minutes and then washed with distilled water. Each sample was divided into four sub-samples (four repetitions) with 25 grains each, which were placed equidistantly in gerbox-type boxes containing a sheet of sterilised filter paper moistened with distilled water.

The boxes were then incubated for seven days at 25 °C in a germination chamber.

After seven days, each seed was checked under a magnifying glass to observe and count the number of seeds that had developed fungal microbiota. To obtain the value of the PFG variable, the number of seeds with fungal growth in each sub-sample was checked, the result was converted into a percentage and the arithmetic mean of the four repetitions was calculated.

The data was analysed using the Assistat software, version 7.7 beta (SILVA e AZEVEDO, 2002) at a 5% probability level. For qualitative variables, means were compared using the Tukey test at a 5% probability level. For quantitative variables, when the treatments were significant, linear or quadratic regression was applied according to what the Assitat software indicated: when both were significant, the one with the highest R^2 was chosen.

RESULTS AND DISCUSSIONS

The summary of the analysis of variance for the variables sprouted grain per ear (GBE), presence of fungus in the ear (PFE) and percentage of fungus in the grains (PFG) are shown in Table 1.

- Sprouted Grains on the Ear (GBE)

Table 1 shows an interaction between the sowing density and nitrogen levels factors for the GBE variable. These results are shown in Figure 2. Significant results are also observed for the harvest time factor, which saw an increase in ears with sprouted grains, from 12.08% in season 1 to 24.58% in season 2 (Table 1).

Figure 2 shows that, according to the regression analysis, for the sowing density of 45,000 plants ha-1, there was a reduction in the GBE variable from 0 to 83 Kg N ha^{-1}, which increased again from this last dose. For the densities of 55,000, 65,000 and 75,000 plants ha^{-1} the GBE, according to the regression, has a maximum point at the respective nitrogen doses of 128.57, 71.43 and 65 Kg N ha^{-1}.

- Presence of fungus on the ears (PFE)

The EFP variable showed no interaction between the factors (Table 1). However, by isolating and carrying out the individual analyses, we can see that there were statistical differences between the harvesting seasons, with season 1 showing a lower percentage of fungi than season 2, with values of 6.25 and 23.75% respectively (Table 1).

Table 1. Summary of the analysis of variance, with the percentages for the variables: grains sprouted on the cob (GBE), presence of fungus on the cobs (PFE) and presence of fungus on the grains (PFG) with the calculated F values, for the interactions and their variations, significance of the linear and quadratic variances and average harvest times of maize plants grown in the off-season (2015) in Dois Vizinhos - Paraná.

	GBE (%)	PEF (%)	PFG (%)
	Harvest Times (CE)		
1	12,08B	6,25 B	2,77 B
2	24,58A	23,75 A	57,79 A
F-value	13,85**	20,26 **	521,45 **
DMS	6,71	7,77	4,82
	Sowing Density (DS) (plants ha)$^{-1}$		
45.000	19,17	13,33	28,50
55.000	20,00	18,33	30,96
65.000	15,00	15,83	28,08
75.000	19,17	12,5	33,58
F-value	0,45	0,46	1,11
F-value Regr. Linear	0,11^{ns}	0,08 ^{ns}	1,32 ^{ns}
F-value Regr. Square	0,25^{ns}	1,15 ^{ns}	0,40 ^{ns}
	Nitrogen Levels (NN) (Kg N ha)$^{-1}$		
0	17,50	20,83	30,50
50	19,17	13,33	30,42
100	20,00	13,33	29,92
150	16,67	12,50	30,29
F-value	0,21	1,01	0,01
F-value Regr. Linear	0,01^{ns}	2,07 ^{ns}	0,01 ^{ns}
F-value Regr. Square	0,55^{ns}	0,74 ^{ns}	0,00 ^{ns}
CE X DS	0,43^{ns}	0,14 ^{ns}	2,02 ^{ns}
EC X NN	0,84^{ns}	2,28 ^{ns}	0,34 ^{ns}
DS X NN	2,74**	0,57 ^{ns}	0,30 ^{ns}
EC X DS X NN	0,66^{ns}	0,78 ^{ns}	0,29 ^{ns}
CV (%)	89,77	126,93	38,98

** and^{ns} : significant at 1% and not significant respectively.

Averages followed by the same capital letter in the column differ by Tukey's test at 1% and 5% probability levels.

EC = harvest time. DS = sowing density, NN = nitrogen level.

Fungus in the Grain (PFG)

Table 1 shows statistical differences between the maize harvest times for the PFG variable, rising from 2.77 to 57.79 per cent from season 1 to season 2.

Table 1 shows that 2.77% of the seeds showed fungal growth in season 1, with this figure rising to 57.79% in season 2. It is worth noting that in season 1, only 12.5% of the subsamples contained at least one seed with fungal microbiota growth, while in season 2 100% of the subsamples showed fungi. Among the fungi found, the genera Aspergillus spp., Fusarium spp. and Penicillium spp. stand out (Figure 3).

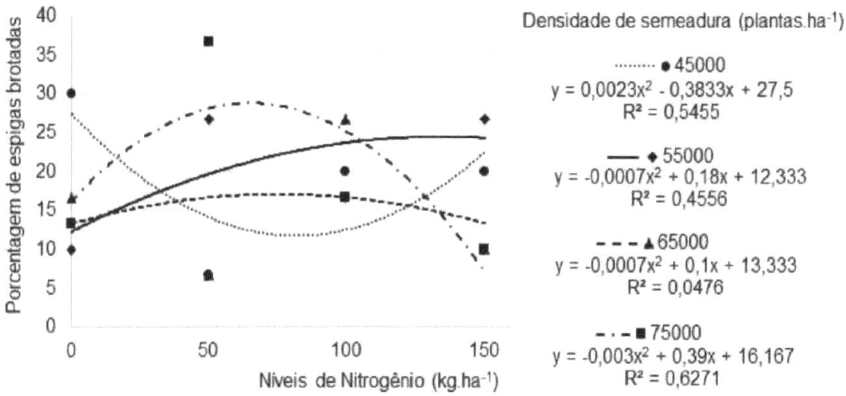

Figure 2 Percentage of sprouted ears, as a function of sowing density versus nitrogen fertiliser levels, of maize grown in the off-season, 2B587Hx (2015) in the municipality of Dois Vizinhos - Paraná.

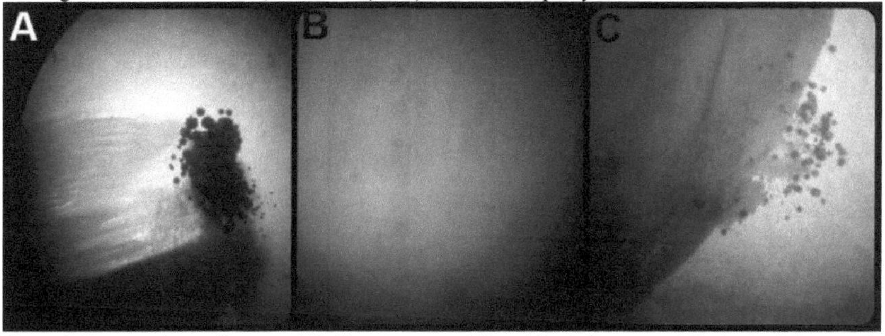

Figure 3 Fungus: Aspergilus spp. (A), Fusarium spp. (B) and Penicilium spp. (C) obtained from maize grains grown in the off-season, 2B587Hx (2015) in the municipality of Dois Vizinhos - Paraná. Photo: Vanderson Vieira Batista.

It can be seen in Table 01 that all the grain health variables worsen as the harvest is delayed, or the longer the maize remains in the field, the greater the incidence of fungi and sprouted kernels.

It was observed that the grains sprouted on the cobs and the presence of fungi were in the upper part of the cob and that few plants had decumbent cobs, suggesting that rainwater infiltrated through the straw of the cob, soaking the upper grains and stimulating them to sprout.

According to Giehl et al. (2011), the stuffing of maize cobs is one of the factors that can directly influence the occurrence of fungi, and the more the cob is exposed to the environment, the greater the chances of contamination with fungal spores. This increase in fungi on the cobs is due to the cobs remaining in the field for a longer period of time. This factor, combined with the reduction in temperatures and increase in humidity, favours the formation of a microclimate in the field that is favourable for the development of fungi (FONTOURA et al. 2006).

A considerable number of ears were found to have fungal growth on the outside of the ear, even before the grains reached physiological maturity. This suggests that the presence of the pathogens in the environment occurred while the plants were still young, contaminating the immature ears, and then these fungi were transferred to the inside of the ear, contaminating the grains as well.

The concern about the presence of these fungi goes beyond the direct loss of grain weight and discounts to producers due to burnt grains, since these fungi are described as having the potential to produce various types of mycotoxins, which are toxic metabolites of a heterogeneous nature and with various pharmacological principles, which can act on the animal organism, impairing its performance and developing serious pathological alterations (SANTIN et al., 2000).

Giehl et al. (2011) assessed the incidence of Penicillium spp. and Fusarium spp. in 8 cultivars of Creole maize and reported a large number of fungi in all the cultivars, with Penicillium spp. being found in 70 to 100% of the samples and Fusarium spp. ranging from 43 to 97% contamination.

The maize reached the point of physiological maturity on 26/05/2015 (season 1) and was harvested on 16/05/2015 (season 2), twenty-one days apart. The fact is that after the physiological ripeness of the maize, which occurs at 30 to 33% grain moisture, moisture loss occurs through evaporation of the water, which is difficult to dissipate into the atmosphere due to the low temperatures and high relative humidity, which are common climatic conditions at this time of year. It so happens that during this period when the maize remains in the field, the conditions of high humidity and low temperatures favour fungal development.

To avoid problems with burnt grains and ear rot, Pinto (2005) makes a series of recommendations, including avoiding high plant density, not delaying harvesting and using maize cultivars with decumbent ears.

The cultivar's ear-stuffing capacity and the decumbent ear characteristic were not assessed. However, at harvest time, it was observed that a low number of plants expressed this characteristic when ripe, which is an important factor that may have contributed to the appearance of fungal microbiota on the grains and ear rot.

CONCLUSIONS

There is a quadratic relationship between nitrogen doses and plant density for the presence of sprouted grains in the ear.

The length of time the maize remains in the field affects the health of the grains, increasing the number of sprouted grains and the incidence of fungal microbiota on the cob and in the grains of maize grown in the off-season in the municipality of Dois Vizinhos - PR.

BIBLIOGRAPHICAL REFERENCES

ALVARES, C. A.; STAPE, J. L.; SENTELHAS, P. C.; MORAES, J. L. G; SPAROVEK, G. Kõpperfs climate classification map for Brazil. Meteorologische Zeitschrift, Stuttgart, v. 22, p.711-728, 2013.

BHERING, S. B., DOS SANTOS, H. G., BOGNOLA, I. A., CÚRCIO, G. R., CARVALHO JUNIOR, W. D., CHAGAS, C. D. S., SILVA, J. D. S. Soil map of the state of Paraná, updated legend. In: Embrapa Florestas-Article in conference proceedings (ALICE). In: BRAZILIAN CONGRESS OF SOIL SCIENCE, 32nd, 2009, Fortaleza. Soil and bioenergy production: perspectives and challenges: proceedings. [Viçosa, MG]: SBCS; Fortaleza: UFC, 2009., 2008.

DOW SEEDS. Seed guide 2013. Dow AgroSciences. 2013.

IAPAR. Agronomic Institute of Paraná. Historical averages at IAPAR stations. Available at: <htpp://www.iapar.br/arquivos/Imagens/monitoramento/Medias_Historicas/Francisco_Bltrao.htm>. Accessed on: 27 October 2014.

FARIAS, A. X.; ROBB, C. F.; BITTENCOURT, A. M.; ANDRESEN, P. M.; CORRÊA, T. B. S.; Endogenous contamination by Aspergillus spp. in post-harvest maize in the state of Paraná. Pesquisa Agropecuária Brasileira. V. 35, n. 3. p. 617-621. March 2000.

FIESP. Federation of Industries of the State of São Paulo. DEAGRO Newsletter. World Corn Harvest 2014/2015 - 6th USDA Survey. October 2014.

FONTOURA, D.; STANGARLIN, J. R.; TRAUTMANN, R. R.; SCHIRMER, R.; SCHWANTES, D. O., ANDREOTTI, M. Influence of plant population on the incidence of stalk diseases in safrinha maize hybrids. Acta Sci. Agron., v. 28, n. 4, p. 545-551. 2006.

FORNASIERI FILHO, D. Manual da cultura do milho. Jabuticabal. Funesp. 2007.

GEBIOMET. Biometeorology study group. Available at: <www.gebiomet.com.br/dowloads.php> Accessed on: 13 April 2016.

GIEHL, J.; REININGER, L. R. S.; NOAL, G.; Evaluation of the relationship between cob stuffing and fungal incidence in Creole maize cultivars. VII Brazilian Congress of Agroecology. Fortaleza - CE. 2011.

MACHADO, J. C.; OLIVEIRA, J. A.; VIEIRA, M. G. G. C.; ALVES, M. C. Control of seed germination in soya beans in a health test by the use of water restriction. Brazilian Seed Journal. Vol. 25, n. 2, p. 77-81, 2003.

PINTO, N. F. J. de A. Burnt grains in maize. Technical circular 66. Embrapa Milho Sorgo, 6p., Jul. 2005.

SANTIN, E.; MAIORKA, A.; ZANELLA, I.; MAGOM, L. Mycotoxins from Fusarium spp. in commercial poultry farming. Ciência Rural, v.31, p.185-190, 2000.

SILVA, F. A. S.; AZEVEDO, C. A. V. Version of the Assistat computer programme for the Windows operating system. Revista Brasileira de Produtos Agroindustriais, Campina Grande, v.4, n.1, p.71-78, 2002.

CHAPTER 9

CHLOROPHYLL CONTENT AND DEVELOPMENT OF MAIZE PLANTS AS A FUNCTION OF THE USE OF HUMIC/FULVIC ACIDS ASSOCIATED OR NOT WITH NITROGEN

Paulo Roberto Rabelo
Eduardo Júnior de Oliveira da Silva
Vanderson Vieira Batista
Karine Fuschter Oligini
Ana Claudia Schllemer dos Santo
Lucas Link
Paulo Fernando Adami

Abstract: Knowing that maize is very dependent on fertilisers, the aim of this study was to quantify the chlorophyll content of the leaves and check the development of the plants at different phenological stages, subjected to different treatments of humic/fulvic acids (biostimulant), associated or not with nitrogen. The study was carried out at the Federal Technological University of Paraná - Dois Vizinhos Campus, in a randomised block design, with five treatments and four replications. The treatments evaluated were: Treatment 1: ^ dose (0.75 L ha^{-1}) of biostimulant at V4 + ^ dose (0.75 L ha^{-1}) at V8 + 170 kg ha^{-1} of N; Treatment 2: recommended dose of biostimulant (1.5 L ha^{-1}) at V4 + 170 Kg ha^{-1} of N; Treatment 3: Recommended dose of the biostimulant (1.5 L ha^{-1}) at V8 + 170 Kg ha^{-1} of N; Treatment 4: Recommended dose of the biostimulant (1.5 L ha^{-1}) at V4; Treatment 5: 170 Kg ha^{-1} of N. The amount of chlorophyll A, B and total was not influenced by the treatments studied in V7, V12 and R1, except for chlorophyll B in R1, where higher values were observed in treatment 5 (16.98) compared to treatment 1 (11.77). As for plant development, at R1 treatment 2, with a height of 221 cm, differed from treatment 4, which had an average plant height of 216 cm. However, the plant height at V7 and V12 and the leaf area of the maize plants at V7, V12 and R1 were similar between the treatments studied.

INTRODUCTION

Due to its high productivity and great importance to Brazilian agribusiness, maize (Zea mays L.) is one of the main crops produced in Brazil, with the country ranking third among the world's largest producers of this cereal (USDA, 2017).

Adequate management practices are fundamental to successful maize cultivation, particularly the use of the no-till system (SPD) and the levels of investment in fertiliser. According to Simione et al. (2017), variations in SPD management have significant effects on the total operating cost of production and the response to fertilisation. The higher the level of technology employed, the higher the grain yield of the maize crop (LUJAN et al., 2015), and Souza and Moreira et al. (2016) point out that nitrogen fertilisation in the form of urea provides greater profitability when growing maize in SPD.

As a grass, maize is very demanding of fertilisers, especially nitrogen (N) (MARTIN et al., 2011). Franco et al. (2013) point out that N is one of the nutrients most required by maize, which is directly related to plant growth and development, as well as increasing grain yields.

According to Franco et al. (2013), plants well nourished with nitrogen have a great capacity to assimilate CO_2 and synthesise carbohydrates in the process of photosynthesis, which consequently also increases the crop's yield components. According to Lima et al. (2001), N is necessary for the synthesis of chlorophyll involved in plant photosynthesis, i.e. N is directly associated with the plant's efficiency in utilising solar energy for its essential functions such as growth and development, so its lack can result in lower biomass accumulation and consequently lower crop yield potential.

Humic and fulvic acids, on the other hand, are naturally released by the decomposition of organic matter and have the potential to stimulate physiological changes in plants, and these changes can help to improve development, which can result in productivity gains in crops (CARON et al., 2015).

Studies have shown the positive effect of humic and fulvic acids on various species, such as the development of tomato seedlings (BERNARDES et al., 2011) and the growth promotion of pineapple seedlings propagated in vitro (BALDOTTO et al., 2009). In maize, Canellas et al. (2008) found that maize seedlings grown in solutions containing humic acids and citric acid modify the exudation profile and that there is an increase in root growth characteristics. Zandonadi et al. (2007) observed that humic acids promote the induction of lateral roots in maize.

According to Nannipieri et al. (1983), humic substances have the capacity to interfere directly in the plant's metabolism, influencing ion transport, respiratory activity, the amount of chlorophyll and the activity of enzymes and nucleic acids.

In this context, the aim of the study is to quantify the chlorophyll content of maize leaves and check the development of maize plants at different phenological stages, subjected to different treatments using humic/fulvic acids, associated or not with nitrogen.

MATERIAL AND METHODS

The study was conducted in the Experimental Area of the Federal Technological University of Paraná - Dois Vizinhos Campus (UTFPR - DV), latitude 25°42'S, longitude 53°08'W and average altitude of 560 metres. The region's climate is Cfa (ALVARES et al., 2013) and the soil is Latossolo Vermelho Distroférrico (BHERING et al., 2008), with flat terrain. The historical average rainfall in the region is approximately 2,000 mm per year, which is distributed throughout the year (IAPAR, 2017).

The study was conducted in a randomised block design with five treatments and four replications. The treatments consisted of different ways of using a commercial product (biostimulant), based on humic/fulvic acids, associated or not with nitrogen. The following treatments were evaluated: Treatment 1: ^ dose (0.75 L ha^{-1}) of the biostimulant at V4 + ^ dose (0.75 L ha^{-1}) at V8 + 170 Kg ha^{-1} of N; Treatment 2: recommended dose of the biostimulant (1.5 L ha^{-1}) at V4 + 170 Kg ha^{-1} of N; Treatment 3: Recommended dose of the biostimulant (1.5 L ha^{-1}) in V8 + 170 Kg ha^{-1} of N; Treatment 4: Recommended dose of the biostimulant (1.5 L ha^{-1}) in V4; Treatment 5: use of 170 Kg ha^{-1} of N.

To carry out the experiment, preparation of the area began 30 days before sowing the maize, with desiccation of the Avena strigosa grown in the area during the winter period. Desiccation was carried out using the chemical herbicide Glyphosate, using 1.5 L ha^{-1} . The study was conducted using the maize hybrid 30F53HY, which was sown on 7 September 2016, in a no-till system, with a density of 80,000 seeds ha^{-1} and a spacing of 45 cm between crop rows. The base fertiliser in the sowing furrow was chemical fertiliser formulated 16-34-00 (N-P2O5-K2O) and 150 kg ha^{-1} of KCl, supplied in the sowing furrow and as top dressing.

After the plants had emerged, at the V2 stage of development, the sampling units (UAs) were marked out (Figure 1 A), which consisted of 36 m^2 , 10 metres long and 3.6 metres wide, covering 8 rows of maize grown at a spacing of 45 cm between rows.

Figure 1. Demarcated sampling units (UAs) (A) and application of the humic/fulvic acid-based biostimulant (B). Source: Author (2018).

The application of the biostimulant was carried out using a CO_2 compressor, coupled to a bar with 4 nozzles, improving the efficiency and distribution of the products

(Figure 1 B). The source of nitrogen (N) used was urea (45% N), and 170 kg ha^{-1} of N was applied manually, by spreading and top dressing, when the maize plants had four fully developed leaves (vegetative stage V4).

Data was collected from plants located in the central five metres of the four central rows of each AU.

Plant development was assessed at three different vegetative stages in V7 (first assessment) and V12 (second assessment) and reproductive stage R1 (third assessment), and the amount of chlorophyll A, chlorophyll B and total chlorophyll were also measured at three different stages: V6 (first assessment), V11 (second assessment) and R1 (third assessment).

Chlorophyll A and B levels were measured on five plants per AU using a DATALOG (Figure 2 A), analysed in the middle of the last fully expanded leaf. The arithmetic mean of the values found was calculated and used for data analysis. Also, by adding up the chlorophyll A and chlorophyll B values found in each AU, the total chlorophyll levels in the respective treatments were established.

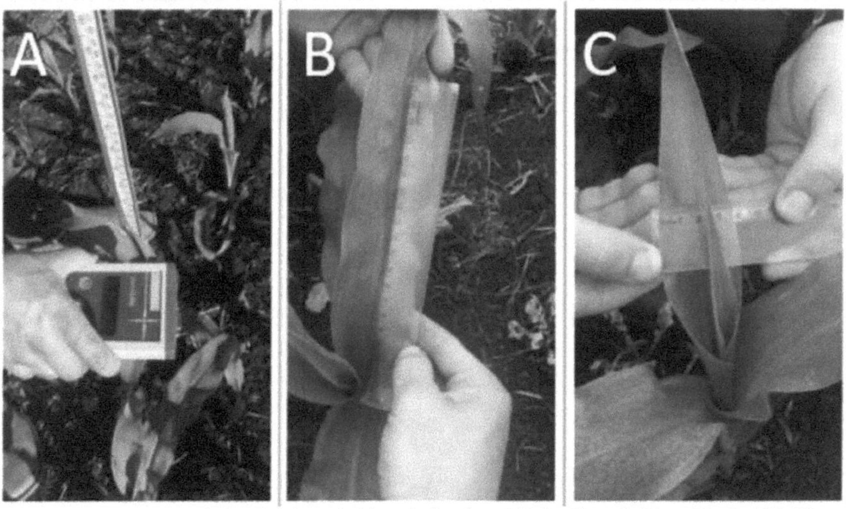

Figure 2: Measurement of plant height and chlorophyll values (A), leaf length (B) and leaf width (C).
Source: Author (2018).

The development of the maize crop was observed on five plants per AU, measuring plant height (cm). These values were obtained using a tape measure, taking into account the value observed between ground level and the point where the leaf blade of the last fully expanded leaf bent.

In addition to plant height, leaf area (cm^2) was also assessed on 3 plants per AU. A ruler was used to measure the length (LF) of the leaf limb of all the leaves (Figure 2 C) and also the width (Figure 2 B) of the leaf (LF) at its centre. The values were tabulated and the leaf area of each leaf was calculated, correcting for the coefficient of 0.7 (leaf area = CF*LF*0.7). The values observed on each plant were added up and the arithmetic mean between plants from each AU was calculated, using the average value observed for data analysis.

Analysis of variance (F-test) was applied and, when significant, Tukey's test was applied at 5% probability, using Assistat 7.7beta software (SILVA e AZEVEDO, 2002).

RESULTS AND DISCUSSIONS

Table 1 shows that there were no differences between the treatments for chlorophyll A, B and total values when the maize was at V7.

Average values of 45.97 (chlorophyll A), 10.93 (chlorophyll B) and 56.87 were reported for total chlorophyll (Table 1). According to Ertani et al. (2011) humic substances derived from the decomposition of lignin stimulated an increase in chlorophyll content. Caron et al. (2015) point out that an increase in chlorophyll content gives the plant a greater capacity to absorb light, which stimulates photosynthetic activity.

As in the first assessment period, no statistical differences were observed between the treatments in the second season (V12), with average values of 49.80 for chlorophyll A, 11.00 for chlorophyll B and 60.81 for total chlorophyll (Table 2).

Table 1. Influence of humic/fulvic acids and nitrogen on the amount of chlorophyll e at development stage V7. Dois Vizinhos - PR (2018).

Treatments Variables	Chlorophyll A	Chlorophyll B	Total chlorophyll
Treatment 1	$47{,}17^{NS}$	$10{,}76^{NS}$	$57{,}93^{NS}$
Treatment 2	47,05	11,37	58,42
Treatment 3	46,10	11,34	57,44
Treatment 4	44,73	11,11	55,84
Treatment 5	44,83	10,09	54,72
Average	45,97	10,93	56,87
dms	5,63	2,35	6,76
CV (%)	5,43	9,56	5,27

NS: Not significant. Treatment 1: $/^{1}_{2}$ dose (0.75 L ha^{-1}) of the biostimulant in V4 + $/^{1}_{2}$ dose (0.75 L ha^{-1}) in V8 + 170 Kg ha^{-1} of N; Treatment 2: recommended dose of the biostimulant (1.5 L ha^{-1}) in V4 + 170 Kg ha^{-1} of N; Treatment 3: recommended dose of the biostimulant (1.5 L ha^{-1}) in V8 + 170 Kg ha^{-1} of N; Treatment 4: recommended dose of the biostimulant (1.5 L ha^{-1}) in V4; Treatment 5: use of 170 Kg ha^{-1} of N.

Table 2. Influence of humic/fulvic acids and nitrogen on the amount of chlorophyll e at development stage V12. Dois Vizinhos - PR (2018).

Treatments / Variables	Chlorophyll A	Chlorophyll B	Total chlorophyll
Treatment 1	$51{,}18^{NS}$	$11{,}15^{NS}$	$62{,}33^{NS}$
Treatment 2	50,29	10,91	61,21
Treatment 3	49,52	10,92	60,44
Treatment 4	48,47	11,02	59,50
Treatment 5	49,54	11,01	60,56
Average	49,80	11,00	60,81
dms	3,38	1,44	4,05
CV (%)	3,02	5,83	2,96

NS: Not significant. Treatment 1: $/^{1}_{2}$ dose (0.75 L ha^{-1}) of the biostimulant in V4 + $/^{1}_{2}$ dose (0.75 L ha^{-1}) in V8 + 170 Kg ha^{-1} of N; Treatment 2: recommended dose of the biostimulant (1.5 L ha^{-1}) in V4 + 170 Kg ha^{-1} of N; Treatment 3: recommended dose of the biostimulant (1.5 L ha^{-1}) in V8 + 170 Kg ha^{-1} of N; Treatment 4: recommended dose of the biostimulant (1.5 L ha^{-1}) in V4; Treatment 5: use of 170 Kg ha^{-1} of N.

In the third evaluation season, there were differences between the treatments in terms of chlorophyll B content (Table 3). Table 3 shows that treatment 5 (170 Kg ha^{-1} of N at V4) had a higher chlorophyll B value (16.98) than treatment 1 (/¹₂ dose V4 + /¹₂ dose at development stage V8 + 170 Kg ha^{-1} of N (V4)), which had chlorophyll B levels of only 11.77 (Table 3). However, the chlorophyll A and total chlorophyll values did not differ statistically and showed average values of 65.25 and 78.79, respectively (Table 3). Although there were differences in the amount of chlorophyll B in the maize plants, this difference was not enough to affect the total amount of chlorophyll, as chlorophyll B represents a small fraction of the total chlorophyll, which is mostly made up of chlorophyll A (Table 3).

Table 3. Influence of humic/fulvic acids and nitrogen on the amount of chlorophyll e at development stage R1. Dois Vizinhos - PR (2018).

Treatments / Variables	Chlorophyll A	Chlorophyll B	Total chlorophyll
Treatment 1	64,00^{NS}	11,77 b	75,78^{NS}
Treatment 2	62,38	12.65 ab	75,03
Treatment 3	66,29	12.69 ab	78,98
Treatment 4	65,99	12.87 ab	78,86
Treatment 5	67,57	16,98 a	85,31
Average	65,25	13,39	78,79
dms	10,30	4,90	11,22
CV (%)	7,01	16,25	6,32

^{NS}Not significant. Averages followed by different letters in the column differ statistically by the Tukey test at 5% probability. NS: Not significant. Treatment 1: /¹₂ dose (0.75 L ha^{-1}) of the biostimulant at V4 + /¹₂ dose (0.75 L ha^{-1}) at V8 + 170 Kg ha^{-1} of N; Treatment 2: recommended dose of the biostimulant (1.5 L ha^{-1}) at V4 + 170 Kg ha^{-1} of N; Treatment 3: recommended dose of the biostimulant (1.5 L ha^{-1}) in V8 + 170 Kg ha^{-1} of N; Treatment 4: recommended dose of the biostimulant (1.5 L ha^{-1}) in V4; Treatment 5: use of 170 Kg ha^{-1} of N.

With no statistical difference for the chlorophyll variables (Table 1, Table 2 and Table 3), except for the amount of chlorophyll B in R1 (Table 3), it is assumed that the maize plants had similar amounts of nitrogen, even in treatment 4, which did not receive nitrogen in the form of urea top dressing. This suggests that the application of the biostimulant was somehow able to meet the N demand of the maize crop.

As for the height of the maize plant, treatment 2 was 221.75 cm taller than treatment 4, which had a plant height of 216.85 cm at stage R1 (Table 4). Despite being statistically different, this difference was only 5 cm, a low value considering the development of the maize crop.

Table 4: Influence of humic/fulvic acids and nitrogen on plant height (PH) (cm) at V6 (PH 1), V12 (PH 2) and R1 (PH 3) and leaf area (LA) (cm^2) at V6 (LA 1), V12 (LA 2) and R1 (LA 3). Dois Vizinhos - PR (2018).

Treatments Variables	AP 1	AF 1	AP 2	AF 2	AP 3	AF 3

Treatment 1	28,50^{NS}	819,21^{NS}	62,95^{NS}	3.227,66^{NS}	217.35 ab	6.579,99^{NS}
Treatment 2	29,00	914,02	62,70	3.521,00	221,75 a	6.654,05
Treatment 3	29,15	964,46	62,35	3.354,92	217.20 ab	6.694,60
Treatment 4	29,97	931,46	63,15	3.066,50	216,85 b	6.883,80
Treatment 5	28,70	922,98	62,75	3.414,43	217.75 ab	6.678,02
Average	29,06	910,42	62,78	3.316,90	218,18	6.698,09
dms	2,31	179,61	6,65	483,34	4,68	852,45
CV (%)	3,52	8,75	4,70	6,46	0,95	5,64

^{NS}Not significant. Averages followed by different letters in the column differ statistically by the Tukey test at 5% probability. NS: Not significant. Treatment 1: $/^{1}_{2}$ dose (0.75 L ha^{-1}) of the biostimulant at V4 + $/^{1}_{2}$ dose (0.75 L ha^{-1}) at V8 + 170 Kg ha^{-1} of N; Treatment 2: recommended dose of the biostimulant (1.5 L ha^{-1}) at V4 + 170 Kg ha^{-1} of N; Treatment 3: recommended dose of the biostimulant (1.5 L ha^{-1}) in V8 + 170 Kg ha^{-1} of N; Treatment 4: recommended dose of the biostimulant (1.5 L ha^{-1}) in V4; Treatment 5: use of 170 Kg ha^{-1} of N.

For the other evaluation periods, plant height was similar between the treatments (Table 4). Table 4 also shows that the leaf area of the maize plants did not differ between the treatments studied in any of the seasons.

As can be seen in Tables 1, 2, 3 and 4, most of the variables showed similar values for the treatments analysed, showing that the biostimulant matches the use of nitrogen, but does not have a superior effect when used in conjunction with maize. Studies carried out by Zandonadi et al. (2007) showed that humic and fulvic acids can result in an increase in the root system, as well as an increase in the biomass of the aerial part.

According to Baldotto and Baldotto (2014), humic acids are known to stimulate plant rooting, as their effect is similar to that of plant hormones of the auxin class, which can promote plant growth, even when supplied in small concentrations. According to Nannipieri et al. (1983), humic substances have the capacity to interfere in the plant's metabolism, influencing ion transport, aiding root respiration, altering the amount of chlorophyll and acting on the activities of enzymes and nucleic acids.

Knowing the various effects that humic/fulvic acids can have on the metabolism and development of plants, coupled with the high level of investment in basic fertiliser when sowing the experimental crop, and understanding that the soil in the experimental area had good fertility, with organic matter values of 41.55 g dm^3 , coupled with favourable climatic factors for the development of the crop during the study, it can be assumed that the maize found favourable conditions for development, and that the biostimulant in these conditions has the potential to make up for the lack of N in top dressing, resulting in a quantity of chlorophyll in the leaf and plant development similar to the use of nitrogen.

It can also be seen that the combination of nitrogen + biostimulant does not increase chlorophyll and does not interfere with the development of the maize crop, compared to the use of both nutrients alone.

CONCLUSION

Under the conditions of the study, the use of the humic/fulvic acid-based biostimulant did not affect the chlorophyll levels, plant height and leaf area of the maize, except at R1 when the chlorophyll B values and plant height were different between the treatments analysed.

BIBLIOGRAPHICAL REFERENCES

ALVARES, C. A.; STAPE, J. L.; SENTELHAS, P. C.; MORAES, J. L. G.; SPAROVEK, G. Kõppen's climate classification map for Brazil. Meteorologische Zeitschrift, Stuttgart, v. 22, p.711-728, 2013.

BALDOTTO, L. E. B, BALDOTTO, M.A., GIRO, V.B., CANELLAS, L.P., OLIVARES, F.L., SMITH, R.B. Performance of 'Vitória' pineapple in response to the application of humic acids during acclimatisation. Revista Brasileira de Ciência do Solo, v. 33, n. 4, 2009.

BALDOTTO, M. A.; BALDOTTO, L. E. B. Humic acids. Revista Ceres, v. 61, p. 856881, 2014.

BERNARDES, J. M., REIS, J. M. R., RODRIGUES, J. F. Effect of humic substance application on tomato seedlings. Global Science and Technology, v. 4, n. 3, 2011.

BHERING, S. B., SANTOS, H. G., BOGNOLA, I. A., CÚRCIO, G.R., CARVALHO JUNIOR, W. D., CHAGAS, C. D. S., SILVA, J. D. S. In: Embrapa Florestas- Article in conference proceedings (ALICE). In: BRAZILIAN CONGRESS OF SOIL SCIENCE, 32nd, 2009, Fortaleza. Soil and bioenergy production: perspectives and challenges: proceedings [Viçosa, MG]: SBCS; Fortaleza: UFC, 2009., 2008.

CANELLAS, L. P., TEIXEIRA JUNIOR, L. R. L., DOBBSS, L. B., SILVA, C. A., MEDICI, L. O., ZANDONADI D. B., FAÇANHA, A. R. Humic acids crossinteractions with root and organic acids. Annals of Applied Biology, 153:157-166. 2008.

CARON, Vanessa C.; GRAÇAS, J. Pereira; CASTRO, PRC. Soil conditioners: humic and fulvic acids. Piracicaba: ESALQ/USP, 2015.

ERTANI, A.; FRANCIOSO, O.; TUGNOLI, V.; RIGHI, V.; NARDI, S.A. Effect of commercial lignosulfonate-humate on Zea mays L. metabolism. Journal of Agricultural and Food Chemistry, Washington, v. 59, n. 22, p. 11940-11948, 2011.

FRANCO, A. A. N.; MARQUES, O. J.; FILHO, P. S. V. Safrinha maize production systems in Paraná. XII National Seminar on Stability and Productivity of Safrinha Maize. Dourados, MS. 2013.

IAPAR. Agronomic Institute of Paraná. Maize crop zoning 2ª harvest in Paraná: maps. Electronic version. Available at :< http://www.iapar.br/modules/conteudo/conteudo.php?conteudo=1089>. Accessed on: 24 October 2017.

LIMA, E. V.; ARAGÃO, C. A.; MORAIS, O. M.; TANAKA, R.; GRASSI FILHO, H. NK fertilisation on the development and concentration of macronutrients in flowering bean. Scientia Agricola, v. 58, n.1, p. 125-129, jan./mar. 2001.

LUJAN, D. W., MULLER, A. L., SIBALDELLI, R. N. R., AMARAL, H. F., FERREIRA, R. C. Influence of Technological Levels on Grain Yield of Different Commercial Maize Hybrids in a Red Dystrophic Oxisol. GLOBAL SCIENCE AND TECHNOLOGY, v. 8, n. 1, 2015.

MARTIN, T. N.; PAVINATO, P. S.; SILVA, M. R.; ORTIZ, S.; BERTONCELI, P. Nutrient flow in conserved forage production ecosystems. In: TURCO, Giselle M. S. Anais. IV Symposium: Production and Utilisation of Conserved Forage. Maringá, p. 173-219, 2011.

NANNIPIERI, P.; MUCCINI, L.; CIARDI, C. Microbial biomass and enzyme activities: production and persistence. Soil Biology and Biochemistry, 15:679- 685, 1983.

SILVA, F. A. S.; AZEVEDO, C. A. V. de. Version of the Assistat computer programme for the Windows operating system. Revista Brasileira de Produtos Agroindustriais, Campina Grande, v.4, n.1, p.71-78, 2002.

SIMIONI, F. J., BARTZ, M. L. C., DO PRADO WILDNER, L., SPAGNOLLO, E., DA

VEIGA, M., BARETTA, D. Indicators of technical and economic efficiency of no-till maize in the state of Santa Catarina, Brazil. Revista Ceres, v. 63, n. 6, 2017.

SOUZA, J. A., MOREIRA, A. Economic viability of nitrogen sources and doses in the cultivation of second crop maize in a no-till system. Revista de Ciências Agrárias/Amazonian Journal of Agricultural and Environmental Sciences, v. 58, n. 3, p. 308-313, 2016.

USDA. United States Department of Agriculture. World Corn Harvest 2015/16 - 14th USDA Survey. DEAGRO Newsletter (FIESP Agribusiness Department) Electronic version, April 2017. Available at: <http://www.fiesp.com.br/indices- pesquisas-e-publicacoes/safra-mundial-de-milho-2/>Accessed on: 22/01/2018.

ZANDONADI, D. B.; CANELLAS, L. P.; FAÇANHA, A. R. Indolacetic and humic acids induce lateral root development through a concerted plasmalemma and tonoplast H+ pumps activation. Planta, Berlin, v. 225, p. 1583-1595, 2007.

MIX
Papier aus verantwortungsvollen Quellen
Paper from responsible sources
FSC® C105338

Printed by Books on Demand GmbH, Norderstedt / Germany